MOLECULAR CONNECTIVITY
IN
CHEMISTRY AND DRUG RESEARCH

MEDICINAL CHEMISTRY

A Series of Monographs

EDITED BY

GEORGE DESTEVENS

*CIBA Pharmaceutical Company, A Division of CIBA Corporation
Summit, New Jersey*

Molecular Connectivity
in
Chemistry and Drug Research

LEMONT B. KIER

Department of Chemistry
Massachusetts College of Pharmacy
Boston, Massachusetts

LOWELL H. HALL

Department of Chemistry
Eastern Nazarene College
Quincy, Massachusetts

ACADEMIC PRESS   New York   San Francisco   London   1976
A Subsidiary of Harcourt Brace Jovanovich, Publishers

NMU LIBRARY

ACADEMIC PRESS, INC.
111 Fifth Avenue, New York, New York 10003

United Kingdom Edition published by
ACADEMIC PRESS, INC. (LONDON) LTD.
24/28 Oval Road, London NW1

Library of Congress Cataloging in Publication Data

Kier, Lemont Burwell.
    Molecular connectivity in chemistry and drug research.

    (Medicinal chemistry, a series of monographs)
    Includes bibliographical references.
    1.   Molecular theory.   I.   Hall, Lowell H.,   Date
joint author.   II.   Title.   III.   Series.
  QD461.K42   541'.22   76-18696
  ISBN 0−12−406560−0

To our fathers
Lemont B. Kier and Lowell H. Hall
for their inspiration, encouragement, and example

CONTENTS

Chapter One   Structure and Properties

Chapter Two   Elements of Graph Theory and Topological Indices

Chapter Three   Molecular Connectivity

Chapter Four   Molecular Properties and Connectivity

Chapter Five   Molar Properties and Molecular Connectivity

Chapter Six   Connectivity and Nonspecific Biological Activity

Chapter Seven   Substituent Group Structure–Activity Relationships

Chapter Eight   Multiple Chi Terms Relating to Biological Activity

PREFACE

The organic or medicinal chemist has, for many years, employed topology in his consideration of molecular structure. A structural formula is, in reality, a topological graph; a skeleton formula is a subgraph; a heterocyclic molecule is depicted with a rooted circuit graph; branched isomers of molecules are distinguished by formulas reflecting different connectivities. There is a well-developed intuition that different molecular structures, described by different topological graphs, have different properties.

Until now this intuition has been qualitative. It is obvious that butane and isobutane structural representations are different. It is not apparent from the structural formula how different they are. Is this difference greater or less than the difference between isopentane and neopentane? In other words, is it possible to assign some numerical value to the graphs of molecules, so that differences in structure could be quantitated? Beyond this, is it possible to differentiate numerically molecular structures sufficiently so that significant correlations are possible with physical, chemical, and biological properties?

These possibilities represent real opportunities to the organic and medicinal chemist in the study of structure–activity relationships. If numerical values could be assigned, or better yet, developed nonempirically, which reflect meaningful aspects of molecular structure, these scientists would have a powerful tool to analyze and predict numerical values of properties of molecules that are of interest to them.

This book describes a new approach to the quantitative evaluation of molecular structure, which we call molecular connectivity. It is a nonempirical derivation of numerical values that encode within them sufficient information to relate to many physicochemical and biological properties. We have discovered these relationships many times.

The method of molecular connectivity is extremely simple, while providing a flexibility to consider important heteroatoms. Furthermore, the method has the inherent ability to describe numerically a molecule at several levels of consideration, each level conveying different information about the connectivity of the molecule. The composite of these extended connectivity values has brought the correlation with some physical properties to a point near the experimental limit of the values.

In this book we will develop the method of molecular connectivity as it has evolved in our laboratory to date. This is followed by a section on the application to physicochemical properties. The next section shows how the method can be applied to structure–activity studies in medicinal chemistry. The final chapter contains some reflections, current challenges, and future areas of investigation of molecular connectivity.

ACKNOWLEDGMENTS

A number of people have contributed significantly to this work and deserve acknowledgment. Our early collaborators, M. Randić and W. J. Murray, made essential contributions during the formative stages of the work. Valuable technical assistance was rendered by J. Fisk, P. Coy, and D. LaLone. Helpful technical discussions were contributed by G. Amidon, A. Cammarata, J. U. Free, R. H. Mann, T. DiPaolo-Chênevert, J. McCloy, W. J. Murray, E. B. Roche, and S. Sickler. Computer assistance was generously supplied by the Eastern Nazarene Computer Center, L. A. Baker, Director, and the Massachusetts College of Pharmacy Computer Center, F. Parmenter, Director. Much of the preliminary draft was typed by D. D. Hall, while the final manuscript was typed by M. L. Kier.

Chapter One

STRUCTURE AND PROPERTIES

In chemistry we seek relationships between the fundamental nature of atoms and molecules and their behavior as expressed in experimental quantities. The concept that molecules consist of atoms bound together into stable, identifiable entities has played a vital role in modern chemistry. Physical properties, stability, reactivity, and other characteristics are described and explained in molecular terms. Much creative effort at many levels of theory has been devoted to the development of methods that relate what we know of structure to what we measure as properties.

At the heart of any science is the awareness that changes in composition or structure lead to changes in properties and function. Chemistry is no exception to this rule. Indeed, we are acutely aware of the profound influences that modest structural variation in molecules has upon physical, chemical, and biological properties. As a consequence, a large part of the study of chemistry is devoted to the subject of the definition of structure.

At this time, quantum mechanics is the ultimate approach to the quantification of molecular structure. Given the coordinates and atomic numbers for a collection of atoms, the Schroedinger equation, in principle, can be solved for the eigenvalues and eigenvectors that describe the energy and electron distribution. The stable arrangement of these atoms in molecular form corresponds to the lowest energy arrangement. Other properties of the molecular aggregate are derived from the wave function and energy.

For example, if we were to conduct a sophisticated quantum mechanical calculation of the eigenvalues and eigenvectors of a molecule containing four carbon and ten hydrogen atoms, we must introduce the

1

numbers and kinds of atoms. Solution would yield the structures of the two most stable isomers, butane and isobutane. Our chemical experience tells us that these are the only stable combinations of these atoms under reasonable conditions.

Our intuition, based on classical notions of valence, would lead us to the same prediction. Normally the chemist could not accurately describe the electronic structure or energies of the two isomers without a quantum mechanical calculation.

There are two levels of structural information concerning a molecule. The complete structure, both electronic and geometrical, is obtained through quantum mechanics. At an intermediate level it is possible to write the structural formulas of isomers based on intuitive notions of chemical bonding. This intermediate level of structural information is the bonding or branching pattern in the molecule. Structural information concerning branching, atom connections, shape, and size can be classified under the general term *topology*.

I.   Structural Influences on Physicochemical Properties

Numerous examples are available illustrating the influence of structure on experimental properties of molecules. Some of these are presented in Table I. It is apparent that each property bears a relationship to the molecular structure, although the nature of this relationship is variable. The molecular weight is strictly additive in terms of the numbers and kinds of atoms in the molecule. Additivity is fundamental to the concept of an homologous series in organic chemistry. The heat of atomization in the hydrocarbon series is additive within experimental error, the increment per methylene unit being 280.03 kcal/mole. Molar volume and molar refraction are also perceived to be additive in this series.

In contrast, pure additivity is not found in this series for the properties of boiling point and specific gravity. In each of these cases, the increment between successive members in the homologous series slowly decreases.

In this hydrocarbon homologous series, as well as in others, such properties as heat of atomization and molar refraction show excellent linear correlation with the number of carbon atoms. In contrast, the properties of boiling point and density show a nonlinear correlation.

The relationship between molecular structure and properties is less direct when we consider molecules that are branched. Table II illustrates some of these properties for the isomeric hexanes. None of the

TABLE I

Structural Influences on Selected Properties of Alkane Homologous Series

Compound	Heat of atomization[a]	Molar refraction[b]	Molar volume[c]	Molecular weight	Refractive index[a]	Boiling point[a]	Specific gravity[a]
Butane	1234.96	—	—	58.13	—	−5.0	—
Pentane	1514.80	25.27	115.22	72.15	1.3575	36.07	0.6262
Hexane	1794.72	29.91	130.68	86.17	1.3749	68.74	0.6594
Heptane	2074.75	34.54	146.52	100.19	1.3876	98.43	0.6838
Octane	2354.86	39.19	162.58	114.21	1.3974	125.67	0.7025
Nonane	2634.76	43.83	178.69	128.23	1.4054	150.81	0.7176
Decane	2914.84	48.47	194.84	142.25	1.4119	174.12	0.7301

[a] Heat of atomization in kcal mol^{-1} taken from Cox and Pilcher [19, Table 34].

[b] Molar refraction R_m calculated as $[(n^2 − 1)/(n^2 + 1)](M/d)$, in cm^3 mol^{-1}, where n is refractive index and d density.

[c] Molar volume V_m calculated as M/d, where M is molecular weight and d density.

[d] Data taken from Handbook of Tables for Organic Compound Identification, CRC Press, Cleveland, Ohio.

TABLE II

Structural Influences on Selected Properties of an Alkane Isomeric Series[a]

Compound	Heat of atomization	Molar refraction	Molar volume	Molecular weight	Refractive index	Boiling point	Specific gravity
n-Hexane	1794.72	29.91	130.68	86.17	1.3749	68.74	0.6594
3-Methylpentane	1795.93	29.80	129.72	86.17	1.3765	63.28	0.6643
2-Methylpentane	1796.57	29.95	131.92	86.17	1.3715	60.27	0.6532
2,3-Dimethylbutane	1797.41	29.82	130.25	86.17	1.3750	57.99	0.6616
2,2-Dimethylbutane	1799.28	29.93	132.73	86.17	1.3687	49.74	0.6492

[a] All definitions and data sources are the same as in Table I.

properties is the same for any two isomers, hence number of atoms in the molecule is insufficient to describe all of the salient features of the structure that govern the magnitude of the property.

In a superficial analysis, we have listed the hexanes in increasing order of branching based on our perceived intuition of this structural characteristic. The boiling points are seen to decrease in this order, while the heats of atomization increase. In contrast, the molar volumes, molar refractions, and specific gravities are all apparently poorly correlated with respect to this intuitive ordering of the molecules. In each case, however, the properties have a different value for each isomer. Properties in this series, therefore, depend on structure, but that structural quantitation is not always predictable from simple intuitive notions of degree of branching.

At this point it is possible to presume that a complete quantum mechanical treatment of a series of molecules may not be necessary to develop enough information about structure to correlate with some physicochemical properties. If a quantification of the topology of molecules, which we call *molecular connectivity*, could carry with it

TABLE III

Physical Properties with Limited Dependence on Topology

Compound	Ionization potential (eV)	Base ionization constant
1-Chlorobutane	10.67	
2-Chlorobutane	10.65	
1-Chloro-2-methylpropane	10.66	
2-Chloro-2-methylpropane	10.66	
1-Aminobutane	8.71	10.61
2-Aminobutane	8.70	—
1-Amino-2-methylpropane	8.70	10.72
2-Methyl-2-aminopropane	8.64	10.68
Ethane	11.5	
Propane	11.1	
Butane	8.64	
Isobutane	9.23	
Pentane	10.35	
Isopentane	10.32	
Neopentane	10.35	

sufficient structural information, a close correlation may be possible. This is the objective of the approach described in this book and termed molecular connectivity [1–4].

Certain properties do not have a strong dependence on molecular topology. Ionization potential I_p, arising primarily from a single Schroedinger equation eigenvalue, may be strongly dependent on the presence of a single structural feature. As shown in Table III, chlorobutanes all have essentially the same value for I_p, whereas saturated noncyclic alkanes show a relation to structure similar to those properties in Table I and II. Base dissociation constants for aminobutanes also reveal weak dependence on topology. The information derived from molecular topology may be insufficient to establish a basis for good correlation to these properties. The more complete quantum mechanical approach is required.

II.   Applications of Structure Definition

The principal value of structural information, whether it is derived from quantum mechanics or from an intermediate topological level, is the explanation and prediction of physical and chemical properties. This approach, generally termed structure–activity relationship (SAR) studies, has found wide application in chemistry in the prediction of both properties and the course of reactions. A classic example is the use of a numerical value, assigned to an atom or a chemical group to predict its electronic influence on another portion of the molecule. This is exemplified by the Hammett linear free energy relationship. The value, designated σ, is derived from relative values of the pK_a's of aromatic acids, substituted on the ring. The ratio of the K_a values is considered to be a measure of varying electronic influences of the ring substituents.

It should be stated here for the sake of rigorous definition that the σ values of Hammett are ratios derived from one property, used to relate influences on other properties. As Norrington has pointed out, this is an example of property–activity relation (PAR) [5]. The σ value is not a structural characterization but a manifestation of the structure.

Biological properties of interest to medicinal chemists, such as relative potency of drugs, also depend on molecular structure. The topological influence is illustrated by some selected data in Table IV. Nonlinear dependence on the number of atoms is frequently observed. Various patterns of dependence on the degree of branching is also typical. The combination of these two factors has rendered difficult the development of relationships between molecular structure and biological activity.

TABLE IV

Structural Influence on Drug Activity of Selected Alcohols

Compound	log MBC (mM)[a]	pC[b]	log(1/c)[c]	pC[d]
Methanol	3.09			
Ethanol	2.75			
Propanol	2.40			
n-Butanol	1.78			
n-Pentanol	1.20			
n-Hexanol	0.56			
n-Heptanol	0.20			
n-Butanol		1.46	1.42	0.87
Isobutanol		1.54	1.35	—
sec-Butanol		1.16	—	0.60
tert-Butanol		0.98	0.89	0.46

[a] The logarithm of the minimum blocking concentration for nonspecific local anesthesia from D. Agin, L. Hersch, and D. Holtzmann, *Proc. Nat. Sci.* **53,** 952 (1965).

[b] The negative logarithm of concentration from C. Hansch and W. J. Dunn, *J. Pharmacol. Sci.* **61,** 1 (1972).

[c] Relative activity for tadpole narcosis from Overton, *Studies on Narcosis,* Fischer, Jena, Germany (1901).

[d] Relative activity on Madison Fungus from R. H. Baechler, *Proc. Am. Wood Preserv. Assoc.* **43,** 94 (1947).

As an approach to this problem, investigators have sought relationships between experimentally observable properties and biological activity. For example, the partition coefficients between oil and water of series of drug molecules have been used to analyze biological activity [6]. Other properties include the Hammett σ term, molar refractivity, and empirical terms depicting steric influences. These efforts have resulted in some good correlations with biological activity but have not truly achieved a structure definition.

Other studies on drug molecules have considered the electronic structures and reactivity indices derived from approximate quantum mechanical calculations [7]. This is an example of SAR. Unfortunately, the approximate nature of these methods, necessary for large molecules, results in a substantial loss in the information content regarding the structure. As a result this approach is still in its promising infancy.

We have seen from our introductory considerations that quantum mechanics, in principle, gives a complete structural description of a molecule. At an intermediate level, we can consider a molecule as an assemblage of atoms, connected by bonds as Lewis proposed [8]. Such an approach expresses the topological structure of the molecule and may be used as a basis for the quantitative characterization of the connectivity of the structure.

III.   The Chemical Bond Model

The basic assumption of the bond model of molecules is that a significant characteristic of the molecule is the manner in which the atoms are associated. The set of connections between pairs of atoms constitutes and includes much basic information about the molecular structure.

The complexities of the numerically expressed information resident in the total wave function and associated eigenvalues are transformed into a topological representation of the molecule, such as a structural formula. Certainly such a transformation causes the loss of some information.

A hierarchy of structural descriptions can be shown for a molecule as in Fig. 1. At the most primitive level A, the information is limited to the types of atoms present in the molecule. From this, we can draw certain generalizations derived from chemical and physical experience with other molecules of this composition. We can say something about the molecule in terms of hydrocarbon properties and chemistry.

At level B of structural description, more information is conveyed. The kinds of atoms and their combining ratio in a molecule are evident from this formula. This conveys information that the molecule is an alkane. Other generalizations about the chemistry and properties can be drawn, beyond what is conveyed from level A information.

At level C, the information conveyed is much greater but a specific molecule is not identified due to isomeric possibilities. Nevertheless, at this level of information we can speak about approximate values of physical properties such as solubility, boiling point, and density, as well as chemical reactivity.

At information level D there is conveyed for the first time the information of how the atoms in the molecule are organized or connected. Level D reveals the topology of two structural isomers not apparent from level C information. Physical and chemical experience can now be applied to derive a large amount of information concerning the properties of these molecules. At this level much chemical intuition used by the scientist is focused.

At level E, the quantum mechanical description, all of the information

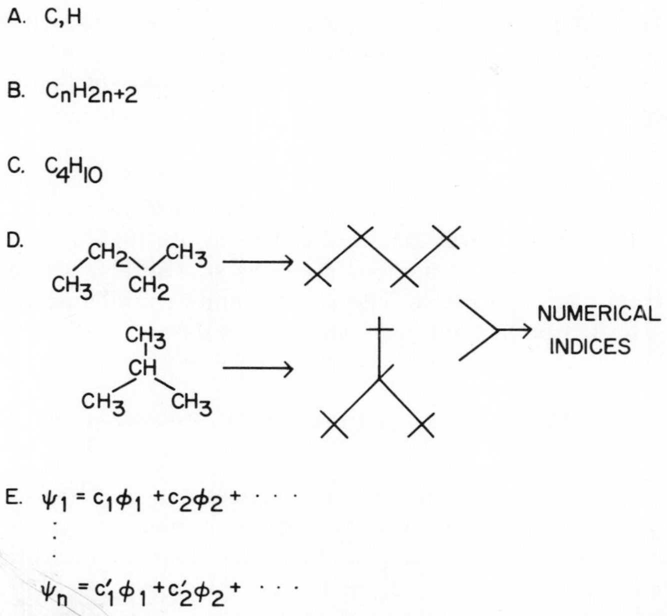

A. C,H

B. C_nH_{2n+2}

C. C_4H_{10}

D.

E. $\psi_1 = c_1\phi_1 + c_2\phi_2 + \cdots$

$$\vdots$$

$\psi_n = c_1'\phi_1 + c_2'\phi_2 + \cdots$

Fig. 1. A hierarchy of structural descriptions from the most primitive at level A to the most complete at level E. Level D represents the topology of the molecule and is the basis for numerical indices of structure.

about the molecule is contained, in principle. For the first time in our present considerations, this information takes numerical form. It includes the probabilities of position and energies of electrons in the molecule.

However, even the prodigious efforts devoted to quantum mechanical calculations over the last two decades have not produced practical methods for predicting properties of large molecules, especially for interacting systems.

There is good reason for the lack of readily usable quantum mechanical methods that yield accurate results for chemical and biological systems. First, the enormous complexity of dealing with all the interactions between particles as required by the Schroedinger equation causes practical problems of computation time. In the second place, a quantity such as total bonding energy is only a tiny fraction of the total calculated energy of a molecular system. A small but acceptable error in terms of the total energy becomes the major portion of the chemical bonding energy. Even for a stable diatomic molecule with a strong bond, such as

N_2 (225 kcal), the total energy of the molecule is more than 100,000 kcal. To ask of a calculation as complex as one based on the Schroedinger equation that the results be better than one part per 100,000 is to ask a great deal. Thus, for these and other reasons, much effort has been directed at relationships above the level E.

At level D, which is the common level of chemical intuition, a very large and useful amount of physical and chemical information can be obtained. It is at this level that we become aware that some properties are directly related to the number of atoms in a series, whereas others deviate from such a relation. The formal numerical characterization of level D information has not been achieved to date.

IV. Additive and Constitutive Properties

Inspection of the relation between structure and properties reveals two general classes of properties. Those properties which may be obtained as a sum of the corresponding values for the constituent parts are called additive. Although molecular mass is the only additive property,* others such as heat of atomization and molar volume are approximately so.

A property that depends heavily on details of the arrangement of the constituent atoms is called constitutive. As illustrated in Table II, boiling point is clearly constitutive. Water solubility of organic compounds is also highly constitutive.

There are many classic cases in chemistry in which the chemical bond model is used as a basis for correlating and predicting physicochemical properties. Values are assigned to atoms, bonds, or groups; these constitutive values are summed in accordance with the molecular topology to yield a value for the whole molecule. All of these methods are based on the assumption of the independence of the bonds.

Perhaps the earliest attempt in this field was the work of Kopp in 1885 [9] on molar volume. Kopp demonstrated the approximate additivity of molar volume, expressed as the ratio of the molecular weight to density, by assigning values to individual atoms. A similar approach was used by Sugden [10] in his work on the parachor.

Molar refraction has been treated in an analogous fashion. Bruhl in 1880 [11] and Eisenlohr in 1910 [12] established tables of atom equivalents of refraction, which could be summed to give a value for a

---

* Because of the relation between mass and energy, based on the theory of relativity and expressed as $E = mc^2$, even molecular mass is not strictly additive.

molecule. In 1921 von Steiger [13] pointed out that refraction is a measure of the deformation of electron distribution. Hence a more rational approach might be made by dissecting the molar refraction into bond contributions. Both Fajans in 1924 [14] and Smyth in 1925 [15] independently established tables of bond values yielding molecular values superior to those based on atom equivalents.

Information based on bond description of structure has been quite successfully used in predicting thermochemical properties. Such methods are developed and described by Pauling [16]. Although a summation of bond energy terms gives a good approximation to heats of formation, branched compounds exhibit significant deviations [17]. It was Fajans [18] who first recognized the influence of nearest neighbors on the heats of formation. Various methods of dealing with thermochemical data have been based on methods of including nearest-neighbor effects [19].

Klages [20] expanded on previous work by introducing terms for heats of combustion and hydrogenation of tertiary and quaternary carbon atoms, as well as for features such as rings. Some improvement is accomplished by this method. Other investigators have introduced additional schemes [21, 22]. Complex sets of correction terms were developed by Tatevskii [23].

An additional step in the development of these methods was made by Franklin [24], who introduced the use of contributions from groups of atoms such as $-CH_2-$, $-CH=CH_2$, and $-CHO$. By this technique Franklin formally accounted for the dependence of each bond on the nature and number of neighboring bonds. Others such as Allen [25] and Laidler [26] have extended these group methods. The most sophisticated version of the group or fragment approach is the work of Zwolinski [27, 28]. Large numbers of various molecular fragments are counted and entered into many parameter equations for the purpose of multiple regression against experimental values of heats of atomization, formation, and vaporization.

In this section we have briefly traced the development of methods for relating molecular structure to properties. A trend is observed in the way of calculating additive and constitutive properties. From early atomic equivalents, to bond additivity, to correction terms, to group or fragment contributions, larger and larger portions of the structure are taken into consideration. There has been a corresponding improvement in calculated results. It would appear that the most useful methods, at a non-quantum mechanical level of calculation, must formally consider the whole molecule. The structural information encoded within a topological representation of a molecule could provide the basis for a useful non-quantum mechanical approach.

V. The Need for Quantitative Molecular Connectivity

The need is evident for a quantitative way to describe molecular structure at the level of its topology. Such a method must develop numerical descriptors encoding within them information relating to the number of atoms and their environment or connectivity. The objective of any method is the development of structural descriptors for correlation with properties dependent on molecular connectivity.

With the preceding discussion in mind, the following criteria may be set forth as guidelines in the evaluation of a method for quantitation of molecular connectivity.

A.   The method should serve as a basis of structural definition with the capacity for wide application to the many physical, chemical, and where possible, biological properties.

B.   The approach should be built fundamentally upon principles of molecular structure rather than upon empirical quantities.

C.   The method should make use of simple computations that are not time consuming, but readily computerized.

D.   The numerical descriptors should be unique for a given structure.

E.   For practical application the method must be sufficiently flexible to handle such structural features as heteroatoms, unsaturation, cyclization, and aromaticity. Such applications should not involve extensive addition of new parameters or features greatly increasing the number of operations or tedium in calculation.

F.   The method should lend itself to amalgamation with certain indices derived from quantum mechanical approaches. Such a combination of topological and molecular orbital indices may provide a powerful tool in the development of SAR.

VI.   Conclusion

It appears that for the present, quantum mechanical methods are not available to calculate accurately properties of large isolated molecules and especially for interacting molecules. Further, it seems clear that when such quantum mechanical approaches become successful, the necessary calculations will be complex and time consuming. For some time to come it is not going to be possible for the practicing chemist to carry out routine quantum mechanical calculations enabling him to predict or correlate solubility, density, boiling point, enzyme inhibition, or narcotic effect.

Yet, if chemistry and biology are to advance in these areas of research, a practical method is needed for the development of relationships between molecular structure and properties. The salient features of structure may be represented by topological means. To the extent that topological characteristics may be numerically expressed, SAR may be advanced. The objective of this book is to describe a method called molecular connectivity, which provides a basis for a topological description of molecular structure in numerical form. In the chapters that follow we will develop the necessary basic tools of graph theory, present the formal method of connectivity, and demonstrate how the method is applied to physical, chemical, and biological properties using numerous examples.

References

1. (a) L. B. Kier, L. H. Hall, W. J. Murray, and M. Randić, *J. Pharm. Sci.* **64**, 1971 (1975).
   (b) L. H. Hall, L. B. Kier, and W. J. Murray, *J. Pharm. Sci.* **64**, 1974 (1975).
   (c) W. J. Murray, L. B. Kier, and L. H. Hall, *J. Pharm. Sci.* **64**, 1978 (1975).
2. (a) L. B. Kier, W. J. Murray, and L. H. Hall, *J. Med. Chem.* **18**, 1272 (1975).
   (b) W. J. Murray, L. B. Kier, and L. H. Hall, *J. Med. Chem.* **19** (1976).
3. (a) L. B. Kier, W. J. Murray, L. H. Hall, and M. Rancić, *J. Pharm. Sci.* **65** (1976).
   (b) L. B. Kier and L. H. Hall, *J. Pharm. Sci.* (in press).
4. L. H. Hall, L. B. Kier, and W. J. Murray, unpublished work.
5. F. E. Norrington, R. M. Hyde, S. G. Williams, and R. Wooten, *J. Med. Chem.* **18**, 6046 (1975).
6. C. Hansch and W. J. Dunn, *J. Pharm. Sci.* **61**, 1 (1972).
7. L. B. Kier, "Molecular Orbital Theory in Drug Research." Academic Press, New York, 1971.
8. G. N. Lewis, *J. Am. Chem. Soc.* **38**, 762 (1916).
9. S. Glasstone, "Textbook of Physical Chemistry," 2nd ed., Chapter 8. Van Nostrand Reinhold, Princeton, New Jersey, 1946.
10. J. Sugden, *J. Chem. Soc.* **125**, 32 (1924).
11. J. W. Bruhl, *Z. Phys. Chem.* **7**(1), 140 (1891).
12. F. Eisenlohr, *Z. Phys. Chem.* **75**, 585 (1910).
13. A. L. von Steiger, *Chem. Ber.* **54**, 1381 (1921).
14. K. Fajans and J. Knorr, *Chem. Ber.* **59**, 249 (1926).
15. C. P. Smyth, *Phil. Mag.* **50**, 361, 715 (1925).
16. L. Pauling, "The Nature of the Chemical Bond," 3rd ed., Chapter 3. Cornell Univ. Press, Ithaca, New York, 1960.
17. C. T. Zahn, *J. Chem. Phys.* **2**, 671 (1934).
18. K. Fajans, *Chem. Ber.* **53**, 643 (1920).
19. J. D. Cox and G. Pilcher, "Thermochemistry of Organic and Organometallic Compounds." Academic Press, New York, 1970.
20. F. Klages, *Chem. Ber.* **82**, 358 (1949).
21. G. S. Parks and H. M. Huffman, "Free Energies of Some Organic Compounds" (ACS Monograph No. 60). Chem. Catalog Co., New York, 1932.

22. (a) S. W. Benson and J. H. Buss, *J. Chem. Phys.* **29,** 546 (1958).
    (b) D. W. van Krevelen and H. A. G. Chermin, *Chem. Eng. Sci.* **1,** 66 (1952).
23. V. M. Tatevskii, V. A. Benderskii, and S. S. Yarovsi, "Rules and Methods for Calculating Physicochemical Properties of Paraffinic Hydrocarbons." Pergamon Press, Oxford, 1961.
24. J. L. Frankin, *Ind. Eng. Chem.* **41,** 1070 (1949).
25. T. L. Allen, *J. Chem. Phys.* **31,** 1039 (1959).
26. K. J. Laidler, *Can. J. Chem.* **34,** 626 (1956).
27. G. R. Somayajulu and B. J. Zwolinski, *Trans. Faraday Soc.* **62,** 2327 (1966).
28. G. R. Somayajulu and B. J. Zwolinski, *Trans. Faraday Soc.* **68,** 1971 (1972).

Chapter Two

ELEMENTS OF GRAPH THEORY AND TOPOLOGICAL INDICES

The chemical bond model may be considered as an example of a more general classification known as a graph [1]. Use of the bond model assumes that significant properties of a molecule may be represented as bonds connecting the atoms in a molecule. A graph is a convenient device for representing the connections in a molecule or the topology of a number of objects that form an aggregate. In chemistry this aggregate is the molecule and the graph is a form of the structural formula. The graphs for butane and isobutane are shown in Fig. 1, Chapter 1, alongside the level D structures.

The word graph, as used in this book, has a different meaning from a plot of data points often commonly referred to as a graph. In our context, a graph may be thought of as a diagram of a molecular structure.

In the sense in which we use it, the topology of a system refers to the connections within the system. In the case of molecules the connections are chemical bonds. In the topological approach, emphasis is given to the fact of connection rather than the type of connection. The chemical graph is then useful as a device for the representation of molecular topology.

Graph theory is extensively employed in the study of patterns, networks, scheduling, electric circuits, and routings as diverse as linen supply and garbage collection [1]. This theory has found significant application in electronics, business, manufacturing, and other areas. Its utility in chemistry has only recently been recognized and will be developed further in this chapter. With this background in mind it is necessary to describe some aspects of graph theory in order to proceed with chemical and biological applications.

In order to make use of some basic ideas and concepts, we will introduce some definitions from graph theory. The reader is advised that many terms used in graph theory arise from commonly used words such as set, path, edge, complete, and regular, to name a few. In their common use these words are imprecise. In graph theory they have very precise meanings. To ensure an understanding of these terms and their interrelations within graph-theoretical concepts, a deliberate reading of the following section is advised.

I.   Definitions and Terms in Graph Theory

A graph is a set of points called *vertices*, which are connected by lines called *edges* (I). Each edge joins two vertices. No edge begins and ends at the same vertex. A graph is symbolized by G and is composed of vertices $V(G)$ and edges $E(G)$. The total graph may be designated $G[V(G), E(G)]$, where $E(G)$ is a finite set of edges and $V(G)$ a finite set of vertices. The symbol n will be used to designate the number of vertices and the symbol m the number of edges.

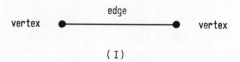

(I)

Illustrations II–V show some definitions in graph theory. The structural formula of methane is only a short step from a graph, as shown in II. The graph for methane is constructed by using dots for vertices

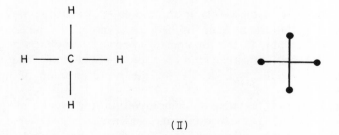

(II)

(atoms) and lines connecting dots for edges (bonds). The graph of ethylene (III) is seen to require two edges connecting a pair of vertices.

These are called *multiple edges* and are drawn as curved or straight lines.

(Ⅲ)

The molecule of 1,1-dimethylcyclohexane (IV) is represented by a graph containing a cyclic structure called a *circuit*. The molecule in V is

(Ⅳ)

(Ⅴ)

an acyclic hydrocarbon. The corresponding graph containing no cycles is called a *tree graph*. A graph in which every pair of vertices is joined by an edge is called a *complete graph* (VI). Such a graph is designated

(Ⅵ)

K_n. From simple principles of combinations it can be shown that K_n possesses $\frac{1}{2}n(n - 1)$ edges. A *star graph* consists of one central vertex connected to three or more surrounding points, which are themselves connected only to the central point. The symbol for a star graph is double subscripted, $K_{1,n}$, where n is the number of vertices connected to the central vertex. The star graphs $K_{1,3}$, $K_{1,4}$, and $K_{1,5}$ are shown in VII.

(VII)

It is customary in chemistry, on occasion, to write only the carbon skeleton of a molecule for which the hydrogen atoms are assumed to be positioned according to the valence of the carbons. A graph of a carbon skeleton is called a *hydrogen-suppressed* graph. The graph for isobutane along with its hydrogen-suppressed graph is shown in VIII.

(VIII)

The *i*th vertex of a graph is designated v_i. An edge may be designated by reference to the vertices at its extremities; these vertices are called *endpoints*. In IX, the edge that connects v_2 and v_3 may be identified as (v_2, v_3) or e_2. In the case of multiple edges (multiple bonds) the designation (v_i, v_j) is ambiguous; hence reference to e_5 or e_6 in IX is preferable to (v_5, v_6).

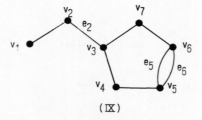

(IX)

The *valence** of vertex v_i is defined as the number of edges that have v_i as one endpoint. The valence is designated $\delta(v_i)$ or simply δ_i. If the number of edges terminating at v_i is j, then $\delta_i = j$. In IX the valence of v_7 is 2, $\delta_7 = 2$, and the valence is three for vertices v_3, v_5, and v_6.

An *edge sequence* in G is defined as a set of edges connected in G and is symbolized by (v_i,v_j), (v_j,v_k), (v_k,v_l), $\ldots$, (v_{m-1},v_m). Each edge terminates on two vertices that are the endpoints of other edges in the sequence, with the exception of terminal vertices. The *length* of a sequence is the number of edges in the sequence. A *path* is an edge sequence in which all edges and vertices occur only once. An edge or vertex that occurs only once in a sequence is said to be *distinct*.

To illustrate, for the hydrogen-suppressed graph of isopentane (X) there are two paths of length 3, designated (v_1,v_2), (v_2,v_4), (v_4,v_5) and (v_3,v_2), (v_2,v_4), (v_4,v_5).

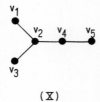

(X)

A *chain* is an edge sequence in which all the edges are distinct. The special case of a chain is a *circuit* in which the first and last vertices are identical.

The designation of a chain in XI is (v_1,v_2), (v_2,v_3), (v_3,v_4), (v_4,v_5), (v_5,v_2), (v_2,v_6). The designation of that part of the graph which is a circuit is (v_2,v_3), (v_3,v_4), (v_4,v_5), (v_5,v_2).

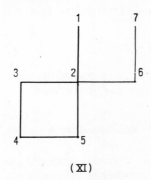

(XI)

---

* Sometimes referred to as the *degree* of the vertex.

A *subgraph* is a designated portion of a graph. Edge sequences are examples of subgraphs. A cyclic graph and some of the subgraphs are shown in XII, subgraphs in dark lines. The second subgraph is a star, $K_{1,3}$, and the third a circuit. A tree graph and some of its subgraphs are shown in XIII.

(XII)

(XIII)

A *spanning tree* is the tree graph obtained from a circuit graph G by removing an edge from each circuit (G must not become disconnected) until G has no circuits. All the spanning trees of a graph are shown in XIV. The notion of a spanning tree finds utility in the consideration of cyclic molecules.

Two graphs G and G' are said to be *isomorphic* if there is a one to one correspondence between vertices in the two graphs as well as with the edges:

$$G(v_i) \equiv G'(v_i), \qquad G(e_i) \equiv G'(e_i)$$

The two graphs in XV are isomorphic although they appear quite different because of the way they are drawn.

II.   Some Relationships in Graph Theory

The practicing organic chemist routinely uses several topological relationships such as the number of isomers possible for a given number of atoms, the number of bonds possible for a given number of atoms in a

(XIV)

(XV)

molecule, and empirical formulas equivalence of olefins and cyclic compounds, to name a few. These have their bases in graph theory and deserve brief mention here.

There are relationships between the number of vertices, the number of edges, and the vertex valencies. A fundamental theorem of graph theory [1] is stated as follows:

If a graph G and n vertices labeled $v_1, v_2, \ldots, v_n$ each with vertex valence $\delta_1, \delta_2, \ldots, \delta_n$ and possessing m edges, then

$$\sum_{i=1}^{n} \delta_i = 2m$$

Thus, the sum of the vertex valencies is an even number.

This theorem has familiar implications to the person who attempts to draw valence structures for organic compounds. One can use the theorem to show that no compound exists for five carbon atoms and nine hydrogen atoms:

$$\sum \delta_i = 5 \times 4 + 9 \times 1 = 29 \neq 2 \times \text{(an integer)}$$

On the other hand five carbons and ten hydrogens must have 15 edges (bonds):

$$\sum \delta_i = 5 \times 4 + 10 \times 1 = 2 \times 15$$

This compound could be a saturated cyclic (XVIII) or a noncyclic compound (XIX) with one double bond (multiple edge).

For hydrocarbons all the vertices of valence greater than one are carbon atoms. When heteroatoms are considered, the vertex corresponding to the different atom may be so identified. Graphs in which such designated vertices are included are called *rooted graphs* and the designated atom is called the root of the graph.

Another relation exists between the number of vertices and the number of edges. This relation depends on the number of cycles in the graph. For a tree graph of n vertices and m edges,

$$m = n - 1$$

For graphs corresponding to solid geometric figures, there is the well-known Euler relation,

$$m = n + c - 2$$

where c is the number of cycles (or faces on the solid figure).

It is now clear that the structural formula qualifies as a graph in the sense of the preceding discussions. See XVI for the graph of ethane

(XVI)

including the hydrogen atoms. Vertices represent atoms and edges stand for bonds. A noncyclic hydrocarbon corresponds to a tree graph as in XVII. A nontree graph represents a cyclic compound as shown in XVIII. An unsaturated compound possesses multiple edges for the multiple bonds as in XIX. Graph XVII is hydrogen suppressed.

(XVII)   (XVIII)   (XIX)

III. The Topological Matrix

In 1874 the mathematician Sylvester [2] showed that a chemical graph could be written as a matrix with entries for edges contributing the nonzero terms in the array. This array is called the *topological* or *adjacency* matrix. Figure 1 shows the transformation from chemical graph to topological matrix. To construct the matrix, the graph is

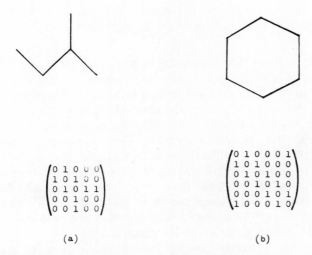

(a)   (b)

Fig. 1. Representation of chemical structures as graphs and matrices.

numbered in any order. The vertex numbers correspond to the row and column designations of the matrix. An entry T_{ij} in the matrix has the value one when there is an edge between vertices i and j; otherwise it is zero.

It can be seen that the matrix is symmetric about the main diagonal. This is true for all topological matrices. It is also apparent that the valence of each vertex is the sum of the unit entries in a row or column corresponding to that particular vertex. Thus, in Fig. 1a the second vertex from the left has two entries in the second row (or column) in the matrix. This is the valence of the vertex. In Fig. 1b all vertices have a valence of 2, corresponding to two unit entries in all rows and columns of its topological matrix.

The topological matrix is a type of mathematical expression that lends itself to computer storage. Thus, we have encountered for the first time in our present discussion a numerical description of a chemical graph or chemical structure.

Algebraic expressions, based on the topological matrix T, for the vertex valence may be formulated as the sum of the entries in a row as

$$\delta_i = \sum_{j=1}^{n} T_{ij}$$

where n is the number of vertices and the order of the T matrix.

In an alternate formulation the diagonal terms in the square of the topological matrix are equal to the valencies

$$\delta_i = \sum_{m=1}^{n} T_{im} T_{mi}$$

The powers T^n of the topological matrix, defined as T multiplied by itself n times, contain additional information as illustrated by the topological matrix of a *digraph*, which is a graph where the edges are directed, in the vectorial sense. The element D_{ij} in the topological matrix of a digraph has the value one when the edge is directed from vertex i to vertex j; otherwise the entries are zero. Thus, the matrix of a digraph is nonsymmetric. Figure 2 illustrates a digraph, its matrix, and the first four powers. Nonzero entries D_{ij}^n in the powers of the digraph matrix indicate the presence of a path from vertex i to vertex j with path length equal to n edges. Thus, there are three paths of length three and one path of length four. All higher power matrices are null and no path of length greater than four exists in the graph.

Powers of the T matrix contain similar information for graphs with nondirected edges as in ordinary (chemical) graphs. Since the edges are

$$D = \begin{pmatrix} 0 & 1 & 0 & 0 & 0 & 0 \\ 0 & 0 & 1 & 0 & 0 & 0 \\ 0 & 0 & 0 & 1 & 1 & 0 \\ 0 & 0 & 0 & 0 & 0 & 0 \\ 0 & 0 & 0 & 0 & 0 & 1 \\ 0 & 0 & 0 & 0 & 0 & 0 \end{pmatrix} \qquad D^2 = \begin{pmatrix} 0 & 0 & 1 & 0 & 0 & 0 \\ 0 & 0 & 0 & 1 & 1 & 0 \\ 0 & 0 & 0 & 0 & 0 & 1 \\ 0 & 0 & 0 & 0 & 0 & 0 \\ 0 & 0 & 0 & 0 & 0 & 0 \\ 0 & 0 & 0 & 0 & 0 & 0 \end{pmatrix}$$

$$D^3 = \begin{pmatrix} 0 & 0 & 0 & 1 & 1 & 0 \\ 0 & 0 & 0 & 0 & 0 & 1 \\ 0 & 0 & 0 & 0 & 0 & 0 \\ 0 & 0 & 0 & 0 & 0 & 0 \\ 0 & 0 & 0 & 0 & 0 & 0 \\ 0 & 0 & 0 & 0 & 0 & 0 \end{pmatrix} \qquad D^4 = \begin{pmatrix} 0 & 0 & 0 & 0 & 0 & 1 \\ 0 & 0 & 0 & 0 & 0 & 0 \\ 0 & 0 & 0 & 0 & 0 & 0 \\ 0 & 0 & 0 & 0 & 0 & 0 \\ 0 & 0 & 0 & 0 & 0 & 0 \\ 0 & 0 & 0 & 0 & 0 & 0 \end{pmatrix}$$

Fig. 2. A digraph and the powers of its topological matrix.

not directed, an edge sequence may traverse a given edge in both directions as illustrated in Fig. 3. Such sequences contain loops. It is clear that matrix methods may be useful in finding all the path subgraphs in a graph. Such applications will be discussed later.

IV. Use of the Topological Matrix in Chemistry

Topology has been useful in predicting the numbers of structural isomers [3]. The topological matrix has served as the basis of calculations in Huckel molecular orbital theory [4]. The early confusion as to why a method so crude could in fact prove so useful in selected areas of application with aromatic molecules has been clarified by Ruedenberg, who showed that a Huckel molecular orbital matrix is a topological matrix [5].

In Huckel molecular orbital theory, each atom in an aromatic or conjugated unsaturated molecule is assumed to be connected to other atoms through the network of the overlapping pi electron domain. This

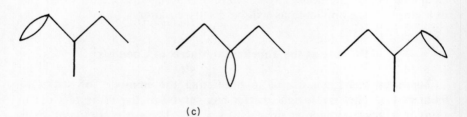

$$T = \begin{pmatrix} 0 & 1 & 0 & 0 & 0 & 0 \\ 1 & 0 & 1 & 0 & 0 & 0 \\ 0 & 1 & 0 & 1 & 1 & 0 \\ 0 & 0 & 1 & 0 & 0 & 0 \\ 0 & 0 & 1 & 0 & 0 & 1 \\ 0 & 0 & 0 & 0 & 1 & 0 \end{pmatrix}$$

$$T^2 = \begin{pmatrix} 1 & 0 & 1 & 0 & 0 & 0 \\ 0 & 2 & 0 & 1 & 1 & 0 \\ 1 & 0 & 3 & 0 & 0 & 1 \\ 0 & 1 & 0 & 1 & 1 & 0 \\ 0 & 1 & 0 & 1 & 2 & 0 \\ 0 & 0 & 1 & 0 & 0 & 1 \end{pmatrix}$$

$$T^3 = \begin{pmatrix} 0 & 2 & 0 & 1 & 1 & 0 \\ 2 & 0 & 4 & 0 & 0 & 1 \\ 0 & 4 & 0 & 3 & 4 & 0 \\ 1 & 0 & 3 & 0 & 0 & 1 \\ 1 & 0 & 4 & 0 & 0 & 2 \\ 0 & 1 & 0 & 1 & 2 & 0 \end{pmatrix}$$

(b)

(c)

Fig. 3. (a) A tree graph, (b) powers of its topological matrix, and (c) selected paths of length five, four and five edges, respectively.

network is treated like the matrix of a graph for the solution of the eigenvalues.

V. Search For a Topological Index

The topological matrix is a mathematical expression that lends itself to manipulations that yield a single number or small sets of numbers that are characteristic of the graph from which they are derived. This leads to the development of a *topological index*. The advantage of such numbers is that they lend themselves to quantitative comparison with physical or chemical properties in a structure–activity study.

The reduction of a graph or a topological matrix to a topological index has been the goal of several investigators. Perhaps the most obvious approach is to expand the matrix T as a determinant to find the eigenvalues of T:

$$T \rightarrow \det|T - xE| \rightarrow \sum_{i=1}^{n} k_i x^{n-i} \rightarrow P(x)$$

Solution for the eigenvalues, for which $-x$ has been added to all diagonal entries, gives a characteristic polynomial $P(x)$. In the notation, E is a unit matrix the same size as T, x the variable, and n the number of vertices in the original graph.

The eigenvalue spectrum of the determinant was thought to be a unique set of numbers and possibly useful as an index. The spectrum, however, has been shown not to be unique [6]. Indeed, Schwenk has shown that almost every tree graph has an isospectral mate [7]. Figure 4 shows several pairs of structures that give identical eigenvalue spectra.

As a result, development of the eigenvalue spectrum does not appear to be fruitful in establishing a unique index. This difficulty has led to a number of different approaches centered on the decomposition of the graph in various ways.

A common feature of the various topological methods is the use of the *hydrogen-suppressed graph*, defined as the graph of the molecule, excluding all hydrogen atoms. Such a graph represents the carbon skeleton of a hydrocarbon. Examples of both structural formulas and the corresponding hydrogen-suppressed graph are given in II–V. The hydrogen-suppressed graph will be used in the remainder of the book.

Most approaches to the development of topological indices involve methods for counting selected topological features. Graph characteristics considered include examples such as the number of pairs of atoms

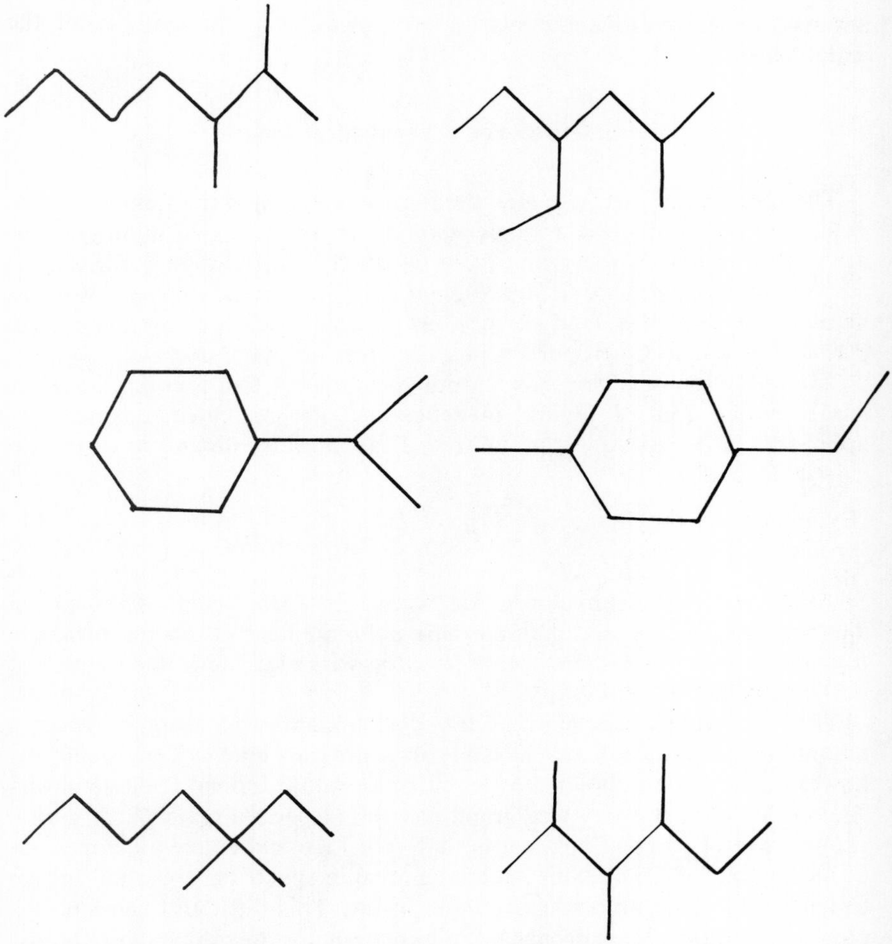

Fig. 4. Pairs of graphs whose eigenvalue spectra are identical, sometimes referred to as cospectral mates.

separated by three bonds. These numbers may be summed to a single integer index or used as coefficients in a polynomial expression.

A. The Wiener Path Number

In 1947, Wiener [8] proposed a topological index that might relate structure to properties of hydrocarbons. The path number w (Wiener number) is defined as the total number of bonds between all pairs of atoms in the (hydrogen-suppressed) graph. The path number may be calculated from the distance matrix D as shown by Hosoya [11]:

$$w = \sum_{i=1}^{n} \sum_{j=i+1}^{n} d_{ij}$$

The elements of D, d_{ij}, are defined as the number of bonds in the shortest path from vertex i to vertex j.

Calculation of the Wiener path number is illustrated in Fig. 5 for 2,3-dimethylpentane. The sum of the upper right triangle of the D matrix element is 46, the value of w. An application to the ordering of boiling points is shown in Table I, column 2 for isomers of hexane. There is a general parallel between w and b.p., but several exceptions stand out.

It can be shown that a simple relationship exists between w and the number of carbon atoms n, only for straight-chain hydrocarbons,

$$w = (n^3 - n)/6$$

Extension of the Wiener approach to complex cyclic graphs would probably require computerized methods for path counting or construction of the D matrix.

B. The Altenburg Polynomial

Altenburg [9] has proposed a polynomial as a characterization of the graph:

$$y = \sum_{i=1}^{N} n_i d_i$$

The count n_i is the number of pairs of atoms separated by d_i bonds. Thus, the expression for 2,3-dimethylpentane is

$$y = 6d_1 + 7d_2 + 6d_3 + 2d_4$$

There is a direct relation of the n_i to the Wiener number, revealed by

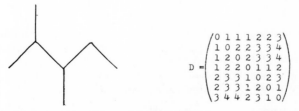

$$D = \begin{pmatrix} 0 & 1 & 1 & 1 & 2 & 2 & 3 \\ 1 & 0 & 2 & 2 & 3 & 3 & 4 \\ 1 & 2 & 0 & 2 & 3 & 3 & 4 \\ 1 & 2 & 2 & 0 & 1 & 1 & 2 \\ 2 & 3 & 3 & 1 & 0 & 2 & 3 \\ 2 & 3 & 3 & 1 & 2 & 0 & 1 \\ 3 & 4 & 4 & 2 & 3 & 1 & 0 \end{pmatrix}$$

Fig. 5. A tree graph and its distance matrix, which may be used in the calculation of the Wiener number and the Altenberg polynomial.

TABLE I

Values of Several Topological Indices Compared with Alkane Boiling Points for Hexanes and Heptanes

Compound	Boiling point[a]	Wiener index[b]	Gordon index[c]	Hosoya index[d]	Gutman index[e]	Randić index[f]
2,2-Dimethylbutane	49.74	31	7	9	24	2.561
2,3-Dimethylbutane	57.99	33	6	10	22	2.643
2-Methylpentane	60.27	35	5	11	20	2.770
3-Methylpentane	63.28	35	5	12	20	2.808
n-Hexane	68.74	38	4	13	18	2.914
2,2,3-Trimethylbutane	80.88	48	9	13	30	2.943
2,2-Dimethylpentane	79.2	50	8	14	28	3.061
3,3-Dimethylpentane	86.03	50	8	16	28	3.121
2,4-Dimethylpentane	80.5	52	7	15	26	3.126
2,3-Dimethylpentane	89.78	52	7	17	26	3.181
2-Methylhexane	90.05	56	6	18	24	3.270
3-Methylhexane	91.85	55	6	19	24	3.308
3-Ethylpentane	93.48	54	6	20	24	3.346
n-Heptane	98.43	60	5	21	22	3.414

[a] Handbook of Tables for Organic Compound Identification, CRC Press, Cleveland, Ohio.

[b] The Wiener index is the sum of the Wiener number w, defined in the text, and p, the number of carbon atoms separated by three bonds. See Wiener [8].

[c] The Gordon index N_2. See text and Gordon and Scantlebury [10].

[d] The Hosoya index Z. See text and Hosoya [11].

[e] See Gutman *et al.* [15].

[f] The Randić branching index. See text and Randić [14].

examination of the distance matrix D. The value of the n_i is the count of the times i appears in the upper triangle of D. The Wiener number can be obtained by letting $d_i = i$ so that

$$w = 6 \times 1 + 7 \times 2 + 6 \times 3 + 2 \times 4 = 46$$

C. Gordon and Scantlebury Index

In an attempt to characterize molecular branching in hydrocarbons explicitly, Gordon and Scantlebury [10] developed a method based on the count of triatomic groups called links. Their index is defined as

$$B_2 = N_2/(n - 1)$$

where n is the number of carbon atoms. The number N_2 is the number of distinct ways in which the graph contains two triatomic groupings or links. Although B_2 does broadly differentiate between major skeletal types, its overall discriminating power is low. Table I lists the values of the index in column 3.

D. The Z Index of Hosoya

Hosoya [11] has introduced an index that bears a formal relationship to the characteristic polynomial of the topological matrix, sometimes referred to as the adjacency matrix of the graph [12]. The index Z is a sum of the nonadjacency numbers $p(k)$:

$$Z = \sum_{k=0}^{m} p(k)$$

The nonadjacency number $p(k)$ is defined as the number of ways in which k bonds may be chosen so that no two of them are connected. It may be shown that

$$p(0) = 1, \qquad p(1) = \text{number of edges in the graph}$$

For the graph in Fig. 6, the nonadjacency numbers are as follows:

$$p(0) = 1, \qquad p(1) = 6, \qquad p(2) = 8, \qquad p(3) = 2, \qquad p(k) = 0, \quad k > 3$$

Hence the Hosoya index Z is 17.

It should be pointed out that the nonadjacency numbers are also the coefficients in the characteristic polynomial for the graph in Fig. 3, without regard to sign:

$$P(X) = X^7 - 6X^5 + 8X^3 - 2X$$

Such a relation exists for all tree graphs but not for graphs containing cycles.

The Z values for hexanes and heptanes are given in Table I, column 4.

E. The Smolenski Additivity Function

The most complex topological method yet advanced is the additivity function of Smolenski [13]. Using the powerful generalizations of graph theory, Smolenski sets forth a procedure purported to deal with additive properties of hydrocarbons.

Smolenski's method involves the decomposition of the graph into sections of various types. The number of each section type then becomes a variable in a multiple regression expression. A linear function

Numbering Scheme	Adjacency Matrix	Binary Equivalent	Decimal Equivalent
	$\begin{pmatrix} 0 & 1 & 0 \\ 1 & 0 & 1 \\ 0 & 1 & 0 \end{pmatrix}$	10, 101, 10	253
	$\begin{pmatrix} 0 & 0 & 1 \\ 0 & 0 & 1 \\ 1 & 1 & 0 \end{pmatrix}$	1, 1, 110	116
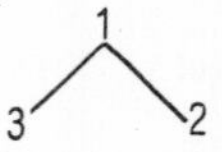	$\begin{pmatrix} 0 & 1 & 1 \\ 1 & 0 & 0 \\ 1 & 0 & 0 \end{pmatrix}$	11, 100, 100	344

Fig. 6. Numbering schemes for the hydrogen-suppressed graph of propane along with the corresponding adjacency (topological) matrices and their interpretation as binary and decimal numbers.

of degree q is defined for graph G:

$$f(G) = a_0 + \sum_{k=1}^{q} \sum_{p=1}^{\sigma_k} a_k^{P_k} |X_k^{P_k}|$$

The coefficients $a_k^{P_k}$ are determined by multiple regression against experimental values for a given property. The quantity σ_k gives the maximum count for a given type of section of k bonds associated with P_k bonds or edges of the graph. The count of sections of each type is given by the symbol $|X_k^{P_k}|$.

An illustration of section types is given in Table II. The small numeral above each carbon atom is the number of carbons joined to that atom, or simply the vertex valence, to use our terminology.

Smolenski has correlated the additivity function with standard heats of formation, ΔH_f^0 for 44 hydrocarbons. The observed and calculated

TABLE II

Graphical Designation of Smolenski Section Types

Section symbol[a]	Section type[a]	Graph fragment[b]
$|X_2^1|$	$\overset{2}{-\text{C}-}$	
$|X_2^2|$	$\overset{3}{-\text{C}-}$	
$|X_2^3|$	$\overset{4}{-\text{C}-}$	
$|X_3^1|$	$\overset{2\;\;2}{-\text{C}-\text{C}-}$	
$|X_3^2|$	$\overset{2\;\;3}{-\text{C}-\text{C}-}$	
$|X_3^3|$	$\overset{2\;\;4}{-\text{C}-\text{C}-}$	
$|X_3^4|$	$\overset{3\;\;3}{-\text{C}-\text{C}-}$	
$|X_3^5|$	$\overset{3\;\;4}{-\text{C}-\text{C}-}$	
$|X_3^6|$	$\overset{4\;\;4}{-\text{C}-\text{C}-}$	

[a] Definitions given in Smolenski [13].
[b] The present authors' interpretation of the Smolenski section type in terms of subgraphs.

results appear in Table III. The best regression equation is

$$\Delta H_f^0 = 25.097|X_2^1| - 10.775|X_2^2| - 6.617|X_2^3| - 20.158|X_3^1| - 10.377|X_3^2|$$
$$-7.038|X_3^3| - 5.601|X_3^4| - 3.950|X_3^5| - 2.824|X_3^6|$$

The fit to observed data is quite good as evidenced by a correlation coefficient $r = 0.9983$ and a standard error $s = 0.610$. Although the standard error is less than twice the experimental error, there are a few notable outstandingly large deviations. There are no branched nonanes in the list and experience indicates that such compounds are more difficult to fit.

F. Randić Branching Index

Randić was interested in the quantitative characterization of the degree of branching [14]. The concept of degree of branching is quite nebulous, with no agreement on criteria for quantitation. As Randić

TABLE III

Alkane Heat of Formation Correlated with the Smolenski Function

Compound	ΔH_f (obs)[a]	ΔH_f (calc)	Residual
Propane	24.82	25.01	−0.19
n-Butane	30.15	29.96	0.19
Isobutane	32.15	32.23	−0.08
n-Pentane	35.00	34.90	0.10
2-Methylbutane	36.92	36.59	0.33
2,2-Dimethylpropane	36.67	39.62	−2.95
n-Hexane	39.96	39.88	0.12
2-Methylpentane	41.66	41.54	0.12
3-Methylpentane	41.02	40.94	0.08
2,3-Dimethylbutane	42.49	42.18	0.31
2,2-Dimethylbutane	44.35	43.62	0.73
n-Heptane	44.89	44.78	0.11
2-Methylhexane	46.60	46.69	0.11
3-Methylhexane	45.96	45.88	0.08
3-Ethylpentane	45.34	45.27	0.06
2,2-Dimethylpentane	49.29	48.55	0.74
3,3-Dimethylpentane	48.17	47.60	0.57
2,2,3-Trimethylbutane	48.96	48.26	0.70
2,3-Dimethylpentane	47.62	46.52	1.10
2,4-Dimethylpentane	48.30	48.17	0.13
n-Octane	49.82	49.72	0.10
2-Methylheptane	51.50	51.43	0.07
3-Methylheptane	50.82	50.83	−0.01
4-Methylheptane	50.69	50.83	−0.14
3-Ethylhexane	50.40	50.23	0.17
2,2-Dimethylhexane	53.71	53.50	0.21
2,3-Dimethylhexane	51.13	51.47	−0.33
2,4-Dimethylhexane	52.44	52.52	−0.08
2,5-Dimethylhexane	53.21	53.12	0.09
3,3-Dimethylhexane	52.61	52.55	0.06
3,4-Dimethylhexane	50.91	50.82	0.09
2-Methyl-3-ethylpentane	50.48	50.87	−0.39
3-Methyl-3-ethylpentane	51.38	51.59	−0.21
2,2,4-Trimethylpentane	53.57	55.20	−1.63
2,2,3-Trimethylpentane	52.61	52.61	0.00
2,3,3-Trimethylpentane	51.73	52.25	−0.52
2,3,4-Trimethylpentane	51.97	52.11	−0.14
2,2,3,3-Tetramethylbutane	53.99	53.93	0.06
n-Nonane	54.74	54.67	0.07
n-Decane	59.67	59.61	0.06
n-Undecane	64.60	64.56	0.04
n-Dodecane	69.52	69.50	0.02
n-Tridecane	74.45	74.44	0.01
n-Tetradecane	79.38	79.38	0.00

[a] All data taken directly from Smolenski [13].

asks, which is more branched, 3-methylhexane or 3-ethylpentane? A definitive answer could lead to a quantitation of structure relating to physical properties, and hence a useful topological index.

The initial stage in the development consisted of creating a scheme for ordering members of an isomeric hydrocarbon series, based on an intuitive idea of molecular branching. The scheme is built around a unique numbering of atoms in a hydrocarbon. The selection of a particular numbering scheme is based on the analysis of the corresponding adjacency matrix. Using propane as an example (Fig. 6), several ways of numbering the vertices are possible. If the rows of the matrix are considered to be binary numbers, three values can be written for each matrix. Interpreting these in decimal notation results in three different numbers. Randić selected the numbering scheme of the hydrocarbon with the lowest decimal number and used that number as an index associated with that particular hydrocarbon.

This procedure provides for a unique way of numbering a hydrocarbon graph and a numerical value that can be associated with it. In more complex graphs, however, the procedure to find the desired numbering scheme is clearly quite complex.

Figure 7 shows the unique numbering scheme for hexane isomers each giving the lowest value of binary (and decimal) interpretation of the adjacency matrix. The molecules are ranked according to increasing value of the binary index. It is noteworthy that the ranking also follows a decreasing order of what we intuitively regard as the degree of branching.

In an attempt to quantify further the degree of branching, Randić directly analyzed elements in the graph. In a graph the connectivities are characterized by adjacency relationships. Beyond this stage lies a consideration of next nearest neighbors. Neighbor relations are embedded in the valencies of the vertices. The edges e of a graph can thus be classified into various types according to the valencies of the endpoints, (δ_a, δ_b). In the case of hydrocarbons there are ten different types of edges since carbon atoms can have hydrogen-suppressed valencies of 1, 2, 3, 4. The edge types are

$$(1,1), \quad (1,2), \quad (1,3), \quad (1,4), \quad (2,2), \quad (2,3), \quad (2,4), \quad (3,3), \quad (3,4), \quad (4,4)$$

If values can be assigned to the individual edge types, then it is possible to sum the values of the edge types in a molecule and to arrive at an index mirroring the order of branching found with the unique numbering scheme, as in Fig. 7.

Based on the unique numbering scheme order of branching, a series of inequalities can be set up for several molecules. Using the hexane

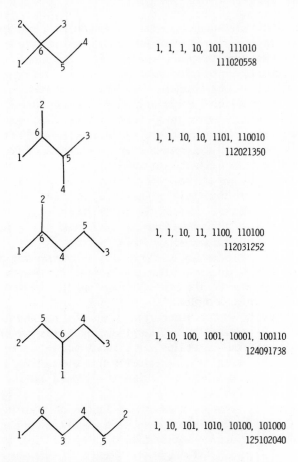

1, 1, 1, 10, 101, 111010
111020558

1, 1, 10, 10, 1101, 110010
112021350

1, 1, 10, 11, 1100, 110100
112031252

1, 10, 100, 1001, 10001, 100110
124091738

1, 10, 101, 1010, 10100, 101000
125102040

Fig. 7. The unique numbering scheme for hexane isomers with the binary interpretation of the adjacency matrix.

isomers in Fig. 7, the inequalities in increasing order are

$$3(1,4) + (1,2) + (2,4) < 4(1,3) + (3,3)$$
$$< 2(1,3) + (2,3) + (2,2) + (1,2) < 2(1,2) + (1,3) + 2(2,3)$$
$$< 2(1,2) + 3(2,2)$$

In higher members of the homologous series, redundancies may arise. Such is the case with 3-methylheptane and 4-methylheptane. Both have the same number of contributing edge types: $2(1,2) + 2(2,2) + 2(2,3) + (1,3)$.

A number of ways may be chosen to satisfy the set of inequalities. The endpoint valencies were considered as a starting point. The reciprocal of the product of vertex valences $1/(\delta_i\delta_j)$ satisfies the inequalities. In order to minimize overlapping of values between different isomer sets, the reciprocal square root was finally chosen. The edge term c_k is thus assigned a value $1/(\delta_i\delta_j)^{1/2}$. These edge values are summed for the entire molecule. Computed branching indices are shown for hexanes and heptanes and compared with boiling points in Table I. The branching index may be expressed in several significant figures and

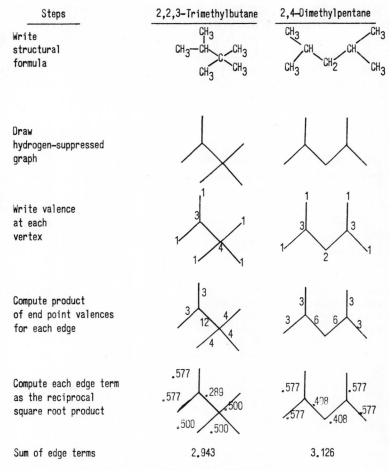

Fig. 8. Detailed procedure for the calculation of the connectivity index $^1\chi$.

shows good correlation with hydrocarbon boiling point. In Fig. 8 the steps in the actual calculation of the index for two graphs are described.

VI. Evaluation of Topological Methods of Structure Representation

A critical test of the methods discussed is their ability to describe structure in such a way as to reveal a relationship with a physical property. Using boiling points in Table IV for a series of hydrocarbons, we have predicted these values using five topological methods. Only the Wiener, Hosoya, and Randić schemes show a reasonable correlation. The Randić scheme is the best of these methods for this property, although for this data set the Wiener index is equally effective. In the beginning of Chapter 3 we will show the clear superiority of the branching index on a broader basis.

We have previously listed several criteria for the evaluation of topological methods for structural representation. An important consideration is the ability to describe accurately structure in relation to

TABLE IV

Alkane Boiling Point Predictions from Several Topological Indices[a]

Compound	Boiling point	Wiener index	Gordon index	Hosoya index	Gutman index	Randić index
2,2-Dimethylbutane	49.74	54.48	79.00	55.19	77.87	51.50
2,3-Dimethylbutane	57.99	57.53	77.24	59.06	74.35	56.08
2-Methylpentane	60.27	60.57	75.48	62.93	70.83	63.17
3-Methylpentane	63.28	60.57	75.48	66.80	70.83	65.29
n-Hexane	68.74	65.14	73.73	70.68	67.32	71.21
2,2,3-Trimethylbutane	80.88	80.37	82.51	70.68	88.42	72.83
2,2-Dimethylpentane	79.20	83.41	80.75	74.55	84.90	79.42
3,3-Dimethylpentane	86.03	83.41	80.75	82.29	84.90	82.76
2,4-Dimethylpentane	80.50	86.46	79.00	78.42	81.38	83.04
2,3-Dimethylpentane	89.78	86.46	79.00	86.17	81.38	86.11
2-Methylhexane	90.05	92.55	77.24	90.04	77.87	91.08
3-Methylhexane	91.85	91.03	77.24	93.91	77.87	93.20
3-Ethylpentane	93.48	89.51	77.24	97.78	77.87	95.33
n-Heptane	98.42	98.64	75.48	101.65	74.35	99.12
Correlation coefficient		0.977	0.160	0.959	0.394	0.979
Standard deviation		3.38	15.68	4.51	14.60	3.26

[a] Correlations are based on the values of the various indices as recorded in Table I.

physical properties. Another important criterion is the potential ability of the method to lend itself to the topological description of molecules other than just hydrocarbons. This would include double bonds and heteroatoms. Within the framework of each method's formalism, it appears that only the Randić scheme is capable of such an extension.

References

1. Selected Reading in Topology and Graph Theory: (a) F. Harary, "Graph Theory." Addison-Wesley, Reading, Massachusetts, 1969.
   (b) R. G. Busacker and T. L. Saaty, "Finite Graphs and Networks." McGraw-Hill, New York, 1965.
   (c) C. Berge, "The Theory of Graphs." Methuen, London, 1962.
   (d) O. Ore, "Theory of Graphs." American Mathematical Society, Providence, Rhode Island, 1962.
   (e) B. Harris (ed.), "Graph Theory and its Applications." Academic Press, New York, 1970.
   (f) R. Wilson, "Introduction to Graph Theory." Academic Press, New York, 1972.
   (g) F. Harary (ed.), "New Directions in the Theory of Graphs." Academic Press, New York, 1973.
   (h) B. H. Arnold, "Intuitive Concepts in Elementary Topology." Prentice-Hall, Englewood Cliffs, New Jersey, 1962.
2. J. J. Sylvester, *Am. J. Math.* **1,** 64 (1874).
3. (a) E. Cayley, *Phil. Mag.* **67,** 444 (1874).
   (b) H. Schiff, *Chem. Ber.* **8,** 1542 (1875).
   (c) H. R. Henze and C. M. Blair, *J. Am. Chem. Soc.* **56,** 157 (1934).
4. J. R. Platt, *in* "Encyclopedia of Physics" (S. Flügge, ed.), Vol. 37, p. 173. Springer-Verlag, Berlin, 1961.
5. K. Ruedenberg, *J. Chem. Phys.* **22,** 1878 (1954).
6. W. C. Herndon, *Tetrahedron Lett.* **8,** 671 (1974).
7. A. Schwenk, *in* "New Directions in the Theory of Graphs" (F. Harary, ed.), p. 275, Academic Press, New York, 1973.
8. (a) H. Wiener, *J. Am. Chem. Soc.* **17,** 2636 (1947).
   (b) H. Wiener, *J. Phys. Chem.* **52,** 425, 1082 (1948).
9. (a) K. Altenburg, *Kolloid-Z.* **178,** 112 (1961).
   (b) K. Altenburg, *Brennst.-Chem.* **47,** 100, 331 (1966).
10. M. Gordon and G. R. Scantlebury, *Trans. Faraday Soc.* **60,** 605 (1964).
11. H. Hosoya, *Bull. Chem. Soc. Japan* **44,** 2332 (1971).
12. D. H. Rouvray, *Am. Sci.* **61,** 729 (1973).
13. E. A. Smolenski, *Russ. J. Phys. Chem.* **38,** 700 (1964).
14. M. Randić, *J. Am. Chem. Soc.* **97,** 6609 (1975).
15. I. Gutman, B. Ruščić, N. Trinajstić, and C. F. Wilcox, *J. Chem. Phys.* **62,** 3399 (1975).

Chapter Three

MOLECULAR CONNECTIVITY

In the last chapter we described several proposed topological indices. Only a few of these appear promising when tested against limited sets of boiling point data for hydrocarbons (see Chapter 2, Table IV). The indices proposed by Wiener [1], Hosoya [2], and Randić [3] appear to have some potential in terms of the correlation coefficient and standard error. Upon examination of a larger set of data (see Table I) for the relationship with solubility, Amidon and Anik [4] have shown that the index described by Randić for hydrocarbons appears appreciably better. A more critical test of these indices using boiling points of a larger set of hydrocarbons leads to the same conclusion, as shown in Table II.

A second advantage of the branching index proposed by Randić lies in the nature of its formulation. All of the other indices described thus far have been designed for hydrocarbons with a constant valence limit of four connections for each atom in the graph. Only for the branching index is the edge count weighted according to the degree of branching. One important criterion of a broadly useful index, as stated in Chapter 2, is the potential of an index formulation to treat heteroatom molecules. This possibility appears evident within the constraint of a limited number of parameters only in the index proposed by Randić. In this scheme the weighting by vertex valence suggests a method for the more detailed consideration of heteroatom connectivity.

A further possibility exists in Randić's formulation for dealing with unsaturated bonds in a molecule. This aspect of the formulation will be considered in a more thorough analysis.

TABLE I

Comparison of Topological Indices for Alkane Water Solubility[a]

Compound	—log S	TSA(A^2)[b]	W[c]	$\sqrt{W}$	Z[d]	log Z	B[e]
Methane	—	152.2	—	—	1	0.00	0.000
Ethane	—	191.7	1	1	2	0.30	1.000
Propane	—	223.5	4	2	3	0.47	1.414
n-Butane	2.63	255.2	10	3.16	5	0.70	1.914
Isobutane	2.55	249.1	6	2.45	4	0.60	1.732
n-Pentane	3.27	287.0	20	4.47	8	0.90	2.414
2-Methylbutane	3.18	274.6	18	4.24	7	0.85	2.270
2,2-Dimethylpropane	3.13	270.1	12	3.46	5	0.70	2.000
Cyclopentane	2.65	—	15	3.87	11	1.04	2.500
n-Hexane	3.95	319.0	35	5.92	13	1.11	2.914
3-Methylpentane	3.83	300.1	31	5.57	12	1.08	2.808
2,2-Dimethylbutane	3.67	290.8	26	5.10	9	0.95	2.561
Cyclohexane	3.18	279.1	27	5.20	18	1.26	3.000
n-Heptane	4.53	351.0	56	7.48	21	1.32	3.414
2,4-Dimethylpentane	4.39	423.7	48	6.93	15	1.18	3.126
Methylcyclohexane	3.85	304.9	42	6.48	26	1.41	3.394
Cycloheptane	3.52	301.9	42	6.48	29	1.46	3.500
n-Octane	5.24	383.0	84	9.17	34	1.53	3.914
1-*cis*-2-Dimethylcyclohexane	4.26	315.5	60	7.75	39	1.59	3.805
Cyclooctane	4.15	322.6	64	8.00	47	1.67	4.000
2,2,4-Trimethylpentane	4.13	338.9	66	8.12	29	1.28	3.417

[a] All data taken from Amidon and Anik [4].
[b] Computed total surface area [4].
[c] Wiener number [1].
[d] Hosoya index [2].
[e] Randić branching index [3].

I.   Analysis of the Randić Scheme as Originally Proposed

An analysis of the scheme proposed by Randić is useful at this point. A number of obvious characteristics of the method should be described, many of which will serve as a basis for in-depth analysis and for the significant expansion of the method introduced since the initial formulation [5–11].

A.   The Hydrogen-Suppressed Graph Representation

The Randić formulation makes use of a hydrogen-suppressed graph in describing hydrocarbons. Some examples are given in Fig. 1. This

TABLE II

Predicted Boiling Points of Alkanes from Several Topological Indices

Compound	Boiling point	Randić index	Boiling point (calc)	Wiener Index	Boiling point (calc)	Hosoya index	Boiling point (calc)
Ethane	−88.63	1.000	−70.55	1	−36.29	2	−26.16
Propane	−42.07	1.414	−40.26	4	−28.76	3	−18.31
n-Butane	−0.50	1.914	−3.674	11	−11.21	5	−2.609
2-Methylpropane	−11.73	1.732	−16.99	9	−16.23	4	−10.46
n-Pentane	36.07	2.414	32.91	22	16.37	8	20.94
2-Methylbutane	27.85	2.270	22.37	20	11.35	7	13.09
2,2-Dimethylpropane	9.5	2.000	2.618	16	1.324	5	−2.609
2,2-Dimethylbutane	49.74	2.561	43.67	31	38.93	9	28.79
2,3-Dimethylbutane	57.99	2.641	49.52	33	43.95	10	36.64
2-Methylpentane	60.27	2.770	58.96	35	48.96	11	44.49
3-Methylpentane	63.28	2.808	61.74	35	48.96	12	52.34
n-Hexane	68.74	2.914	69.50	38	56.48	13	60.19
2,2,3-Trimethylbutane	80.88	2.943	71.62	48	81.56	13	60.19
2,2-Dimethylpentane	79.20	3.061	80.25	50	86.57	14	68.04
3,3-Dimethylpentane	86.03	3.121	84.64	50	86.57	16	83.74
2,4-Dimethylpentane	80.50	3.121	84.64	52	91.59	15	75.89
2,3-Dimethylpentane	89.78	3.181	89.03	52	91.59	17	91.59
2-Methylhexane	90.05	3.270	95.55	56	101.6	18	99.44
3-Methylhexane	91.85	3.303	97.96	55	99.11	19	107.3
3-Ethylpentane	93.48	3.346	101.1	54	96.60	20	115.1
n-Heptane	98.42	3.414	106.1	60	111.6	21	123.0
Correlation coefficient			0.991		0.948		0.913
Standard deviation			6.74		16.4		21.0

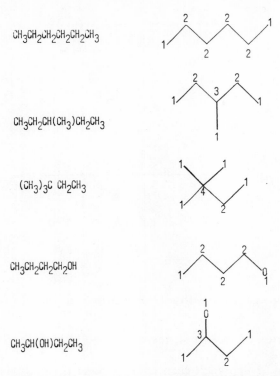

CH₃CH₂CH₂CH₂CH₂CH₃

CH₃CH₂CH(CH₃)CH₂CH₃

(CH₃)₃C CH₂CH₃

CH₃CH₂CH₂CH₂OH

CH₃CH(OH)CH₂CH₃

Fig. 1. Hydrogen-suppressed graphs of molecules showing simple vertex connectivities.

presented no complications since only alkanes, with each carbon having a chemical valence of four, were originally considered.

B. Classification of Edge Types

The differentiation of edge or bond types is based on the designation of each edge or bond. The valence of the endpoints for each edge (or bond) is the source of the numerical value assigned to the edge, c_k. Based on the alkane graphs there are ten types of edges, consistent with chemical experience. This array of bond types is shown in Fig. 2. From inspection of appropriate graphs some logical relationships can be observed. As previously stated in Chapter 2, the sum of the vertex valencies is equal to twice the number of edges. In tree graphs for each vertex of valence 3, there must be at least three vertices with valence 1.

	1	2	3	4
1	1,1	1,2	1,3	1,4
2		2,2	2,3	2,4
3			3,3	3,4
4				4,4

	1	2	3	4
1	1.000	0.707	0.577	0.500
2		0.500	0.408	0.353
3			0.333	0.289
4				0.250

Fig. 2. Array of bond types (upper) and numerical values of bond connectivities (lower).

For each vertex of valence 4, there must be at least four vertices with valence 1.

In the Randić approach each edge or bond is described numerically with a single number, the reciprocal square root of the product of the endpoint valencies. This leads to a spectrum of bond values as shown in Fig. 3. We may consider this in another way. Increasing the valence of one endpoint in the computation of an edge c_k value leads to the relationships shown in Fig. 3.

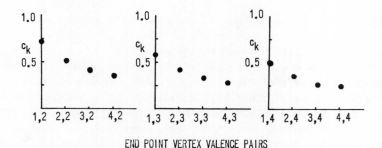

END POINT VERTEX VALENCE PAIRS

Fig. 3. Spectrum of bond connectivities for disubstituted (left), trisubstituted (center), and tetrasubstituted (right) vertices in hydrogen-suppressed graphs.

Two points are obvious from Figs. 2 and 3. First, the c_k values for edge type (2,2) and (1,4) are redundant. Second, the regular increase in valence of one endpoint leads to a nonlinear decrease in the corresponding c_k value. However, for each decrement in c_k value for one edge, there are increased c_k values for other edges, since increasing the valence at a vertex introduces more endpoints with valence 1. The overall effect of increasing valencies of vertices while holding the number of vertices constant is the decrease in the sum of the c_k values or the numerical value of the topological index. Thus, the index for a branched alkane is always less than that for the straight-chain isomer.

It is also apparent that when one vertex valence is changed, as in the case of branching at that point, more than one c_k value is changed. Those values changed are for the edges involved in branching at that point.

The spectrum of index values is generally characteristic of the number of atoms in the isomer series. As the number of atoms increases, the range of values between the straight-chain isomer and the isomer with the lowest topological index steadily increases. This effect is shown in Fig. 4. An overlapping of the range of index values occurs in the octanes in one case, where a highly branched octane has a lower index than does normal heptane. This effect is compounded with higher homologs.

C. Problems Inherent in the Randić Formulation

Several situations are not directly considered in the original formulation by Randić. One of these is the handling of a graph for an unsaturated molecule. In the graph of an alkene, using this formulation, it is not clear what hydrogen suppression in the graph implies nor is it stated how a valence should be written or an evaluation of edges made.

A second problem is the treatment of a cyclic molecule. The graph is easily written but the number of bond terms is always one more (for a monocyclic) than the corresponding straight-chain isomer. What then is the significance of the index?

A third situation not covered in the Randić formalism is the treatment of heteroatoms in molecules. No suggestion is given as to how a different index value is to be computed for the heteroatom-containing molecules. This is an essential consideration if a topological index of value to the chemist and biologist is to be developed.

As an approach to the discussion of the refinement and extension of this topological scheme, it is first desirable to explore in some detail the formalism of the molecular connectivity method.

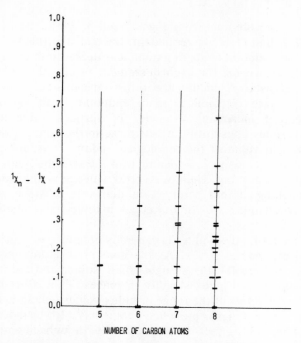

Fig. 4. Spectrum of $^1\chi$ values for the graphs of alkanes of 5 to 8 carbon atoms. Entries are plotted as the numerical difference between the branched value $^1\chi$ and the normal alkane value $^1\chi_n$.

II.   Formal Exposition of the Connectivity Method

We will now set forth the basic principles of the method by which we quantitatively relate molecular topology to properties. Molecular structure is expressed topologically by the hydrogen-suppressed graph. For heteroatomic organic molecules the corresponding graph is rooted. Further, we shall use the *valence-weighted graph* G_v, the hydrogen-suppressed graph bearing the vertex valencies as illustrated in Fig. 1.

Our *first basic assumption* is that structural information necessary for expressing a satisfactory quantitative relationship between the structure of organic molecules and many of their properties is contained in the valence-weighted graph G_v for the molecule. It should be noted that information concerning the contributions of the hydrogen atoms is implicit in this graphical formulation. For alkanes there exists a direct relation between the vertex valence δ_i and the number of hydrogen atoms implied at vertex i, h_i:

$$\delta_i = 4 - h_i$$

The number 4 may represent the valence or the number of valence electrons for the carbon atom. We will discuss this formula later.

In general, properties depend on molecular connectivity as expressed by the graph. The *second basic assumption* we make is that there exists a functional relation between the property and the connectivity characteristics of the graph.

For practical purposes we shall take the functional form of the relation as a sum of terms, each of which depends in a linear manner on graphical characteristics. We define then the *connectivity function* $C(\chi)$ for property p:

$$C(\chi) = b_0 + \sum_{m,t} b_t(m) \, {}^m\chi_t$$

The set of quantities $b_t(m)$ depend on the property and may be determined by multiple regression with experimental data or calculated by a theory or model of the property. In a multiple-regression study, experimental values of a property are regressed against $C(\chi)$. It should be pointed out that the connectivity function is neither an infinite series nor the power series expansion of a function. The highest order term is limited by the number of edges in G_v.

The ${}^m\chi_t$ are terms defined for a subgraph of type t containing m edges, connected in G_v. Disconnected subgraphs are not considered. The order of the subgraph is defined as m. Subgraphs may be conveniently classified into four types. Type 1 ($t = $ P) are called *path* terms and consist of subgraphs whose subgraph valencies are no greater than 2. As defined in Chapter 2, a *path* is an edge sequence with all distinct edges (I). Type 2 ($t = $ C) are called *cluster* terms and consist of subgraphs whose subgraph valencies include at least one 3 or 4 but do not include

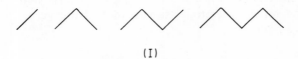

(I)

2. A star subgraph K_{1n} is a special case of type two (II). Type 3 ($t = $ PC) are called *path/cluster* terms whose subgraph valencies must include 2 in addition to 3 and/or 4 (III). Type 4 ($t = $ CH) are called *chain* terms, edge sequences containing at least one cycle. A *circuit* is a special case of a chain (IV).

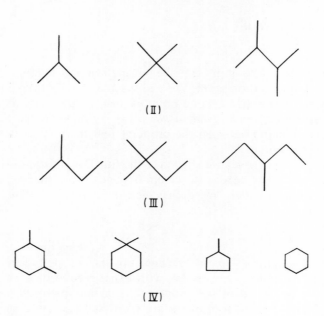

(II)

(III)

(IV)

The connectivity indices $^m\chi_t$ are evaluated as a sum of terms over all the distinct connected subgraphs:

$$^m\chi_t = \sum_{j=1}^{n_m} \, ^mS_j$$

The quantity n_m is the number of type t subgraphs of order m. The subgraph term ^mS_j is a quantity calculated for each subgraph and defined in the next paragraph. All the distinct subgraphs are shown for isopentane in Fig. 5.

In molecular connectivity theory, the most critical aspect of the relation between molecular topology and properties is the dependence on vertex valence. Our *third basic assumption* states that the subgraph terms depend on the reciprocal square root of the vertex valence in a multiplicative manner:

$$^mS_j = \prod_{i=1}^{m+1} (\delta_i)_j^{-1/2}$$

where j denotes the particular set of edges that constitute the subgraph. The number of subgraph valencies multiplied together depends on subgraph type. Chain subgraphs are defined by m vertices, all others by $m + 1$ vertices, where m is the subgraph order.

Thus, the connectivity indices $^m\chi_t$ are valence-weighted counts of connected subgraphs. The weighting scheme introduces the vertex

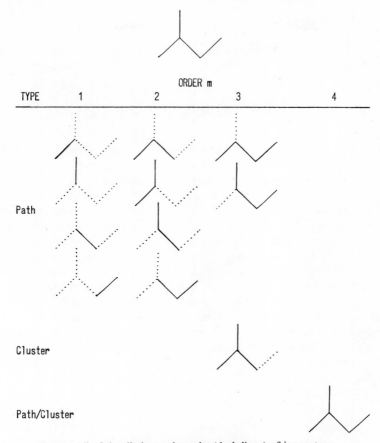

Fig. 5. All of the distinct subgraphs (dark lines) of isopentane.

valence into the relation between (significant) topological features and properties of the molecules. Valence weighting is clearly a key feature in the molecular connectivity method.*

III. Enumeration and Evaluation of $^m\chi_t$ Indices

In order to reveal the nature of the chi terms, we will describe their enumeration and evaluation in detail in this section. This discussion with accompanying tables will clearly show how to calculate $^m\chi_t$ terms.

* The significance of the weighting of subgraph counts by vertex valence and the functional form of the weighting is discussed in Chapter 10.

A.   The Zero-Order Term $^0\chi$

A zero-order subgraph consists of a single vertex, that is, zero edges. Thus, $^0\chi$ is the sum of the reciprocal square root valencies for the graph:

$$^0\chi = \sum_{i=1}^{n} \delta_i^{-1/2}$$

There is, of course, only one type of vertex, represented by its valence, δ. Thus, there is only one type of zero-order subgraph.

Table III shows values for the $^0\chi$ in column 1 for hydrocarbons propane through the isomeric hexanes. Similar information is presented for substituted cyclohexanes in Table IV. For homologous series, $^0\chi$ increases by $(2)^{-1/2}$ ($=0.707$) with the addition of a methylene group. Branching also increases $^0\chi$. Although the $\delta^{-1/2}$ values at the branch point decrease with branching, the number of terminal groups of valence 1 also increases. This increase offsets the decrease at a branch point.

In the calculation of $^0\chi$ no adjacency relations are used; only the number and type of branch points are included. The term $^0\chi$ is a simple composite measure of the number and type of branch points. Hence, numerous redundancies occur. The term $^0\chi$ is the same for 2-methylpentane and 3-methylpentane. Likewise, 3-ethylpentane as well as 2-, 3-, and 4-methylhexanes have identical $^0\chi$ values. Of the disubstituted cyclohexanes in Table IV, 1,2-, 1,3-, and 1,4-dimethylcyclohexane have the same value.

B.   The First-Order Term $^1\chi$

The $^1\chi$ term is a sum over all the edges in the graph, weighted by the reciprocal square root valencies. There is only one type of graph edge. Thus, there is only one type of subgraph of order one:

$$^1\chi = \sum_{s=1}^{N_e} (\delta_i\delta_j)_s^{-1/2}$$

where edge e_s terminates on vertices v_i and v_j. The number of edges in the graph is N_e. It is evident that $^1\chi$ is the branching index suggested by Randić.

Because the topological matrix (or adjacency matrix), designated T, has entries 0 or 1 and expresses both edge count and location, $^1\chi$ can be

TABLE III

Connectivity Indices ${}^{m}\chi_{t}$ for Alkane Graphs: Propane through Isomeric Hexanes

Compound	t	Graph	n	${}^{0}\chi$	${}^{1}\chi$	${}^{2}\chi$	${}^{3}\chi$	${}^{4}\chi$	${}^{5}\chi$
Propane	P		3	2.707	1.414	0.707			
n-Butane	P		4	3.414	1.914	1.000	0.500		
Isobutane	P		4	3.577	1.732	1.732	0.000		
	C						0.577		
n-Pentane	P		5	4.121	2.414	1.354	0.707	0.354	
Isopentane	P		5	4.284	2.270	1.802	0.816	0.000	
	C						0.408	0.000	
	PC							0.408	
2,2-Dimethylpropane	P		5	4.500	2.000	3.000	0.000	0.000	0.250
	C						2.000	0.500	0.000
n-Hexane	P		6	4.828	2.914	1.707	0.957	0.500	0.000
3-Methylpentane	P		6	4.992	2.808	1.922	1.394	0.289	0.000
	C						0.289	0.000	0.289
	PC							0.577	
2-Methylpentane	P		6	4.992	2.770	2.182	0.866	0.577	0.000
	C						0.408	0.000	0.000
	PC							0.289	0.289
2,3-Dimethylbutane	P		6	5.154	2.643	2.488	1.333	0.000	0.000
	C						0.666	0.000	0.333
	PC							1.333	0.000
2,2-Dimethylbutane	P		6	5.207	2.561	2.914	1.061	0.000	0.000
	C						1.561	0.354	0.000
	PC							1.061	0.354

TABLE IV

Connectivity Indices $^m\chi_t$ for Selected Cycloalkanes: Substituted Cyclohexanes

Compound	Graph	t	$^0\chi$	$^1\chi$	$^2\chi$	$^3\chi$	$^4\chi$	$^5\chi$	$^6\chi$	$^7\chi$	$^8\chi$
Cyclohexane		P	4.243	3.000	2.121	1.500	1.061	0.750	0.000		
		CH							0.125		
Methylcyclohexane		P		3.394	2.743	1.894	1.307	0.901	0.204	0.000	
		C				0.289	0.000	0.000	0.000	0.000	
		PC					0.408	0.433	0.000	0.000	
		CH							0.102	0.102	
Ethylcyclohexane		P	5.820	3.932	2.912	2.302	1.595	1.105	0.349	0.000	0.000
		C				0.204	0.000	0.000	0.000	0.000	0.000
		PC					0.493	0.595	0.306	0.000	0.000
		CH							0.102	0.072	0.072
1,1-Dimethylcyclohexane		P	6.036	3.707	3.725	2.207	1.509	1.030	0.354	0.000	0.000
		C				1.207	0.250	0.000	0.000	0.000	0.000
		PC					1.207	1.457	0.625	0.000	0.000
		CH							0.088	0.177	0.088
1,2-Dimethylcyclohexane		P	5.983	3.805	3.239	2.540	1.502	1.024	0.333	0.083	0.000
		C				0.471	0.000	0.167	0.000	0.083	0.000
		PC					1.138	1.040	0.471	0.000	0.000
		CH							0.083	0.167	0.083
1,3-Dimethylcyclohexane		P	5.983	3.788	3.377	2.199	1.737	0.971	0.451	0.000	0.000
		C				0.577	0.000	0.000	0.000	0.000	0.000
		PC					0.742	1.093	0.354	0.000	0.000
		CH							0.083	0.167	0.083
1,4-Dimethylcyclohexane		P	5.983	3.788	3.365	2.305	1.427	1.305	0.333	0.000	0.000
		C				0.577	0.000	0.000	0.000	0.000	0.000
		PC					0.817	0.760	0.471	0.000	0.000
		CH							0.083	0.167	0.083

written in terms of its elements, t_{ij}:

$$^1\chi = \sum_{i=1}^{n} \sum_{j>i}^{n} (\delta_i\delta_j)^{-1/2} t_{ij}$$

The number of vertices is designated by n.

Figure 8 of Chapter 2 shows the individual bond terms in $^1\chi$ for two examples. Values for $^1\chi$ are tabulated in Tables III and IV for a variety of compounds. Because nearest-neighbor adjacency relations enter the calculations for $^1\chi$, some of the redundancies described for $^0\chi$ do not appear in $^1\chi$. Yet, 3- and 4-methylheptane have the same $^1\chi$ as do 1,3- and 1,4-dimethylcyclohexane. Hence, $^1\chi$ possesses greater isomer discriminating power than does $^0\chi$ but is not a unique descriptor.

For other information characterizing the numerical nature of the first-order term refer to Section I of this chapter. It should be noted that for a homologous series $^1\chi$ increases by $\frac{1}{2}$ with each additional methylene group.

C. The Second-Order Term $^2\chi$

Subgraphs of order two contain pairs of connected, that is, adjacent edges and are of one type only, the path type ($t = $ P). Each term contains a reciprocal square root product of the graph valencies for the three vertices defining a distinct subgraph:

$$^2\chi = \sum_{s=1}^{n_m} (\delta_i\delta_j\delta_k)_s^{-1/2}$$

where n_m is the number of subgraphs with two adjacent edges and s identifies a particular subgraph.

The count of subgraph terms of all types is given for isomeric heptanes in Table V. Figure 5 shows all subgraphs terms for isopentane. The number of subgraph terms of order m for linear trees is given as $n_m = n - m$. The number of two-edge subgraphs increases with branching (Table V) in a smooth manner. We will not attempt to discuss here enumeration methods for subgraphs. Later we will refer to an algorithm for enumeration and evaluation of the $^m\chi_t$ terms.

An examination of Tables III and IV reveals that no redundancies exist in $^2\chi$ for the compounds cited. We expect that for graphs of organic molecules, this may be a valid generalization.

D. The Third-Order Term $^3\chi$

In $^3\chi$ we meet the second and third types of subgraphs for the first time in the connectivity function. Path, cluster, and chain terms are

TABLE V

Number of Subgraph Terms for Heptanes: Path, Cluster, and Path/Cluster Types

Graph	$^1\chi$	$^2\chi$	$^3\chi$			$^4\chi$			$^5\chi$			$^6\chi$			Total		
	P	P	P	C	PC	P	C	PC	P	C	PC	P	C	PC	P	C	PC
	6	5	4	0	3	3	0	0	2	0	0	1	0	1	21	0	0
	6	6	6	1	3	3	0	3	0	0	2	0	0	1	21	1	6
	6	6	5	1	3	3	0	2	1	0	2	0	0	0	21	1	4
	6	6	4	1	3	3	0	1	2	0	1	0	0	1	21	1	3
	6	7	6	2	2	2	0	5	0	1	3	0	0	1	21	3	9
	6	7	4	2	4	4	0	2	0	0	4	0	0	1	21	2	7
	6	8	6	4	1	1	1	6	0	0	4	0	0	1	21	5	11
	6	8	4	4	3	3	1	3	0	0	4	0	0	1	21	5	8
	6	9	6	5	0	0	1	9	0	3	2	0	1	0	21	10	11

encountered. Only the star graph $K_{1,3}$ occurs in the cluster term. The triangle subgraph is the only chain subgraph in the third order. Thus, $^3\chi$ may include three types of terms, each evaluated as

$$^3\chi_t = \sum_{s=1}^{n_m} (\delta_i\delta_j\delta_k\delta_l)_s^{-1/2}$$

where n_m is the corresponding number of each type of subgraph and s designates a particular subgraph formed by the edges between the four vertices v_i, v_j, v_k, and v_l. The chain term $^3\chi_{CH}$ is defined by only three vertices.

Various aspects of the numerical nature of $^3\chi$ are shown in Tables III and IV. No redundancies appear for the compounds cited.

Counting the number of subgraphs of each type appears to be a nontrivial matter for $^3\chi$ terms and higher, especially when the number of vertices is large and the branching is complex. Each vertex of valence 3 contributes one $K_{1,3}$ star and each vertex of valence 4 contributes four $K_{1,3}$ stars. Each circuit of three vertices contributes one chain term. However, the number of path terms varies considerably with branching and cyclization as shown in Tables V and VI.

E. The Fourth-Order Subgraph Term $^4\chi$ and Higher Order Terms

In $^4\chi$ we must consider for the first time a case in which all types of subgraph terms are possible: path, cluster, path/cluster, and chain. The $^4\chi$ summation may include all types but is not explicitly written here as such:

$$^4\chi_t = \sum_{s=1}^{n_m} (\delta_i\delta_j\delta_k\delta_l\delta_p)_s^{-1/2} = \sum_{s=1}^{n_m} \prod_i^{m+1} (\delta_i)_s^{-1/2}$$

where n_m is the number of connected subgraphs of type t with four edges. Numerical characterization of $^4\chi_t$ is shown in Tables III and IV. Higher order terms are calculated in a manner analogous to the $^4\chi_t$ terms, including m edges in each case.

F. Generalization on Connectivity Indices

An examination of Tables III and IV along with Appendix A reveals certain general features about the set of indices for each graph. The first observation we make concerns the possible uniqueness of the set of chi values. Since we have observed no counterexamples for graphs of organic molecules, we will state the following *basic conjecture on graph representation uniqueness*:

TABLE VI

The Number of Path Terms in Selected Cyclics

Compound	Graph	m^a								Total
		1	2	3	4	5	6	7	8	
Cyclohexane		6	6	6	6	6				30
Methylcyclohexane		7	8	8	8	8	2			41
Ethylcyclohexane		8	9	10	10	10	4	2		53
1,1-Dimethylcyclohexane		8	11	10	10	10	4	0		53
1,2-Dimethylcyclohexane		8	10	11	10	10	4	1		54
1,3-Dimethylcyclohexane		8	10	10	11	10	5	0		54
1,4-Dimethylcyclohexane		8	10	10	10	12	4	0		54
Propylcyclohexane		9	10	11	12	12	6	5	2	67
Isopropylcyclohexane		9	11	12	12	12	6	4	0	66
1-Methyl-1-ethylcyclohexane		9	12	13	12	12	6	2	0	66
1-Methyl-2-ethylcyclohexane		9	11	13	13	12	6	3	1	68
1-Methyl-3-ethylcyclohexane		9	11	12	13	13	7	3	0	68
1-Methyl-4-ethylcyclohexane		9	11	12	12	14	8	2	0	68
1,2,3-Trimethylcyclohexane		9	12	14	13	12	7	2	0	69
1,2,4-Trimethylcyclohexane		9	12	13	13	14	7	1	0	69
1,3,5-Trimethylcyclohexane		9	12	12	15	12	9	0	0	69

[a] The number of edges defined in the symbol $^m\chi_t$.

The complete set of chi terms $^m\chi_t$ is a unique characterization of a graph (for graphs related to chemical structures). The total chi terms $^m\chi$ may also be unique.

From an examination of Table VII it is clear that for tree graphs similar information is contained in $^0\chi$, $^1\chi$, and $^2\chi$. Although $^0\chi$ and $^2\chi$

TABLE VII

Heptane Isomer Ranking by Chi Terms for $^0\chi$, $^1\chi$, and $^2\chi$ [a]

$^0\chi$	$^1\chi$	$^2\chi$

[a] $^0\chi$ and $^2\chi$ in increasing order, $^1\chi$ in decreasing order.

increase with branching and $^1\chi$ decreases, the ranking (in inverted order) for $^0\chi$ and $^2\chi$ is nearly the same as for $^1\chi$. Only one inversion of order occurs in $^2\chi$ with respect to the other two terms. However, the quantitative rankings and the numerical spread in values is different for the three terms.

Isomer rankings by higher order terms reveal a significantly different picture. Table VIII presents such information for isomeric noncyclic heptanes. Although the first three terms appear to describe the overall

TABLE VIII

Heptane Isomer Ranking by Chi Terms for $^3\chi$ and $^4\chi$ for Path Terms and Total Terms in Increasing Order

$^3\chi_P$	$^4\chi_P$	$^3\chi_T$	$^4\chi_T$

degree of branching, $^3\chi$ and $^4\chi$ do not, when only the path terms are considered (Table VIII, left side). In fact, n-heptane is flanked by monomethyl-substituted hexanes toward the center of the distribution, with highly branched isomers, 2,4-dimethylpentane, 3,3-dimethylpentane, and 2,2,3-trimethylbutane, at the extremes. It appears then that structural features are exalted or minimized in $^3\chi$ and $^4\chi$ in a different manner than in $^1\chi$ and $^2\chi$.

This quality of unlike ranking by $^1\chi$ and $^3\chi$ provides the basis for good correlation with density in hydrocarbons. Neither $^1\chi$ nor $^3\chi$ properly ranks the densities in heptanes. However, when used as two variables in multiple regression, the resulting equation does rank the heptane densities correctly [9]. There is only one exception where two experimental densities are very close in value. This relationship will be discussed in detail in Chapter 5.

As the degree of branching increases in a tree graph for a given number of vertices, the length of the longest path decreases. Hence, high-order $^m\chi_P$ terms are zero for highly branched trees but nonzero for linear trees. The highest order nonzero path term for a linear tree is $m = n - 1$.

For a cyclic graph with n vertices in the cycle, there are n $^m\chi_P$ terms contributed by the cycle alone, up to order $m = n - 1$. The nth-order term becomes a circuit term. Branching on the cycle adds additional path, cluster, and chain terms.

A comparison of the set of χ terms between tree and cyclic graphs containing the same number of vertices reveals some further characteristics about the connectivity function. Some sample information is given in Table IX for six-, seven-, and eight-vertex graphs. It is clear that in every case (except $^0\chi$) the terms are larger for cyclics because the cycle permits more path terms in $^m\chi_P$. This effect is more pronounced for larger m values. The $^0\chi$ values are smaller for cyclics due to the decreased number of terminal groups of valence one in cyclic graphs. In addition, the cyclic graphs shown possess one circuit term, $^6\chi_{CH}$. Thus, it appears that the complete set of chi terms contains information that numerically discriminates between cyclic and noncyclic graphs. This is an important consideration for a formalism that represents properties of chemical compounds.

When all the $^m\chi_t$ terms of a given order m are summed over all types (path, cluster, path/cluster, chain) the resulting term $^m\chi$ is called the total chi term of order m. Ranking of heptane isomers by total $^3\chi$ and $^4\chi$ is shown on the right-hand side of Table VIII. There is a general similarity to the rankings by the lower order terms shown on Table VII. It would appear that this set of total chi terms is a set of numbers that rank

TABLE IX

Comparison of Chi Values for Tree and Cyclic Graphs with Equal Numbers of Vertices and Similar Branching Patterns

Graph	n	$^0\chi$	$^1\chi$	$^2\chi$	$^3\chi_P$	$^4\chi_P$	$^5\chi_P$	$^6\chi_P$	$^6\chi_{CH}$
	6	4.828	2.914	1.707	0.957	0.500	0.500	0.000	
	6	4.243	3.000	2.121	1.500	1.060	0.750	0.000	0.125
	7	5.699	3.270	2.536	1.135	0.612	0.408	0.000	
	7	5.122	3.394	2.743	1.894	1.307	0.901	0.204	0.102
	8	6.621	3.621	3.268	1.883	0.854	0.177	0.000	
	8	6.621	3.561	3.664	1.280	0.707	0.530	0.000	
	8	6.036	3.707	3.725	2.207	1.509	1.030	0.352	0.088
	8	6.569	3.681	3.010	1.882	0.789	0.333	0.000	
	8	6.569	3.719	2.771	2.259	0.805	0.167	0.000	
	8	5.983	3.805	3.239	2.540	1.502	1.024	0.417	0.083
	8	6.569	3.664	3.143	1.571	0.971	0.333	0.000	
	8	5.983	3.788	3.377	2.199	1.737	0.971	0.451	0.083
	8	6.569	3.626	3.365	1.321	0.666	0.666	0.000	
	8	5.983	3.788	3.365	2.305	1.427	1.305	0.333	0.083

isomers according to the degree of branching in a similar but not identical fashion to $^1\chi$.

The number of subgraph terms for each type is presented for noncyclic heptanes for each subgraph order in Table V. It appears that the total number of path terms is structurally invariant for tree graphs of a given number of vertices. This number is equal to $n(n - 1)/2$ or one-half the product of edges and vertices. This number is 21 for noncyclic heptanes, including the number of first-order terms. There appears to be no similar relationship for graphs with circuits. Reference to Table VI shows the total number of path terms for substituted cyclohexanes. This number is nearly constant but depends on the number of branch points, both on the cycle and in side chains.

It is evident that the counting of subgraph terms is a nontrivial matter for graphs of typical organic molecules. Also, the vertices of each subgraph must be identified so that the graph valencies may be entered into the calculation. Hence, an efficient algorithm must be developed for counting, identifying, and evaluating each distinct subgraph connectivity term ^mS_j. Such a method is presented in the next section.

G. An Algorithm for Subgraph Enumeration, Evaluation, and Type Classification

The calculation of the connectivity function subgraph terms may be conveniently considered in four steps for graph G_v with n vertices v_1, v_2, v_3, v_4, . . . , v_n and associated valencies δ_i, N_e edges, for all orders m of subgraphs.

    1. Generation of all possible sets of $m + 1$ vertices from the set $V(G)$ of n vertices in G for chi term $^m\chi_t$.

    2. Determination that the set of subgraph edges $E(G)$, designated subset s, is connected; disconnected sets are no longer considered.

    3. Classification as to subgraph type: path, cluster, path/cluster, and chain on the basis of the set of subgraph valencies δ'

    4. Calculation of subgraph term ^mS_j as

$$^mS_j = \prod_{i=1}^{m+1} (\delta_i)_j^{-1/2}$$

where s indicates the set of vertices in the designated subgraph. The remaining appropriate summations may then be made.

In step 1 the generation of all possible sets of $m + 1$ objects from a set of n objects $(m < n)$ is a well-known problem in permutations and combinations. Standard algorithms are available for such computations.

Subgraph connections may be determined from the entries in the subgraph adjacency matrix A obtained directly from the adjacency matrix for the entire graph. Several illustrations are shown in Table X for 1,1-dimethylcyclohexane. Connectedness can be determined by the subgraph edge count E_s:

$$E_s = \frac{1}{2} \sum_{i=1}^{n} \sum_{j=1}^{n} A_{ij}$$

The following conditions apply for the usual graphs obtained from organic molecules:

a.   A connected tree subgraph, $E_s = m$.
b.   A disconnected subgraph, $E_s < m$.
c.   A cyclic subgraph, $E_s > m$.

An inspection of the subgraph valencies δ_i' reveals the number of 1's, 2's, 3's, and 4's. Hence, from the definition of the subgraph types classification (step 3) is easily accomplished.

Vertex valencies are identified by vertex numbering in the original graph. The final computation of the $^m\chi_t$ terms may be performed by summing ^mS_j terms.

For cyclic subgraphs one difficulty is encountered. A given set of $m + 1$ vertices, defining a circuit, possess $m + 1$ distinct subgraphs of orders 2, 3, 4, . . . , m, which are not recognized by the above procedure. A modification specific for cyclic graphs must therefore be used for the calculations on cyclic structures.

H.   Truncation of the Connectivity Function

In principle, the connectivity function $C(\chi)$ may consist of a large number of terms when all chi terms are considered. For a typical heptane there may be 10–20 nonzero terms, depending on cyclization and degree of branching. The use of such a large number of terms as independent variables in multiple regression analysis, for example, would not always be satisfactory.

However, the connectivity function may be truncated or modified in a number of ways as outlined here:

1.   By using only one type of chi term: for example, only path terms.
2.   By limiting the highest order used.
3.   By assigning the same value of coefficient $b_t(m)$ to all subgraph types. This amounts to using $^m\chi$, the total term, rather than the several types.
4.   By using combinations of 1, 2, and 3.

TABLE X

Illustration of Subgraph Evaluation and Enumeration Algorithm

Example Graph	Topological Matrix
	$\begin{pmatrix} 0 & 1 & 1 & 1 & 1 & 0 & 0 & 0 \\ 1 & 0 & 0 & 0 & 0 & 0 & 0 & 0 \\ 1 & 0 & 0 & 0 & 0 & 0 & 0 & 0 \\ 1 & 0 & 0 & 0 & 0 & 1 & 0 & 0 \\ 1 & 0 & 0 & 0 & 0 & 0 & 1 & 0 \\ 0 & 0 & 0 & 1 & 0 & 0 & 0 & 1 \\ 0 & 0 & 0 & 0 & 1 & 0 & 0 & 1 \\ 0 & 0 & 0 & 0 & 0 & 1 & 0 & 0 \end{pmatrix}$

Vertex Number Sets	Subgraph Matrix	v_i^s	Subgraph	Type
1 2 4 6	$\begin{pmatrix} 0 & 1 & 1 & 0 \\ 1 & 0 & 0 & 0 \\ 1 & 0 & 0 & 1 \\ 0 & 0 & 1 & 0 \end{pmatrix}$	2 1 2 1		Path
	$E_s = 3$			
1 2 3 4 5	$\begin{pmatrix} 0 & 1 & 1 & 1 & 1 \\ 1 & 0 & 0 & 0 & 0 \\ 1 & 0 & 0 & 0 & 0 \\ 1 & 0 & 0 & 0 & 0 \\ 1 & 0 & 0 & 0 & 0 \end{pmatrix}$	4 1 1 1 1		Cluster (star)
	$E_s = 4$			
1 2 3 4 6	$\begin{pmatrix} 0 & 1 & 1 & 1 & 0 \\ 1 & 0 & 0 & 0 & 0 \\ 1 & 0 & 0 & 0 & 0 \\ 1 & 0 & 0 & 0 & 1 \\ 0 & 0 & 0 & 1 & 0 \end{pmatrix}$	3 1 1 2 1		Path/cluster
	$E_s = 4$			
1 4 5 6 7 8	$\begin{pmatrix} 0 & 1 & 1 & 0 & 0 & 0 \\ 1 & 0 & 0 & 1 & 0 & 0 \\ 1 & 0 & 0 & 0 & 1 & 0 \\ 0 & 1 & 0 & 0 & 0 & 1 \\ 0 & 0 & 1 & 0 & 0 & 1 \\ 0 & 0 & 0 & 1 & 1 & 0 \end{pmatrix}$	2 2 2 2 2 2		Circuit
	$E_s = 6$			

Since various properties may depend in various ways on structural features represented by chi terms, statistical analysis may reveal those selected terms which are necessary for satisfactory quantitative relationships. In a series of papers [8–10] it has been shown that the use of a single term or only two or three terms yields very useful results. Such applications are presented in the following sections of this chapter along with other amplifications of the connectivity method. Specific application to physical, chemical, and biological properties are discussed in the remaining chapters.

IV.   Further Development of the Connectivity Method

A.   Heteroatomic Molecules

Practical application of a connectivity index depends heavily on the ability to deal with molecules containing heteroatoms. A heteroatom in a molecule such as diethylether is isoelectronic with a methylene group. Nitrogen, oxygen, and carbon in the first periodic row all share this relationship:

$$CH_3\text{—}CH_2\text{—}\ddot{C}\text{:} \quad (3H) \qquad CH_3\text{—}CH_2\text{—}\ddot{N}\text{:} \quad (2H) \qquad CH_3\text{—}CH_2\text{—}\ddot{O}\text{:} \quad (1H)$$

As a first-level approach to considering the connectivity in a heteroatom molecule we have written the molecular graph and assigned vertex valencies based on the number of first-row atoms bonded to each vertex [5]. Examples are shown in Figure 6. This level of heteroatom treatment leads to a value based strictly on connectivity.

Values of $^1\chi$ calculated in this manner for a series of primary alcohols (Table XII) are shown to correlate closely with several physical properties. A problem arises, however, when primary, secondary, and tertiary alcohols are considered in the same study. Examples 2 and 4 in Fig. 6 lead to identical graphs and values of $^m\chi_t$, although the physical property values are different. Clearly a modification must be introduced to cancel this kind of redundancy.

1.   Empirical Treatment

Two general approaches have been introduced to reflect the different situations presented by heteroatoms in a molecule. One method is to characterize specifically the hydroxyl groups in a series of mixed classes of alcohols in a study of boiling points and solubilities [6]. The $^1\chi$ was calculated for the alkyl portion of the molecule and a second regression variable introduced for the hydroxyl group, the c_k term from the appropriate term in $^1\chi$. This additional term in the regression equation

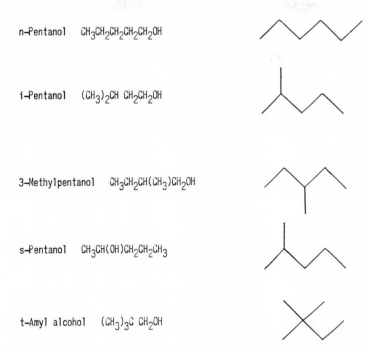

n–Pentanol $CH_3CH_2CH_2CH_2CH_2OH$

i–Pentanol $(CH_3)_2CH\ CH_2CH_2OH$

3–Methylpentanol $CH_3CH_2CH(CH_3)CH_2OH$

s–Pentanol $CH_3CH(OH)CH_2CH_2CH_3$

t–Amyl alcohol $(CH_3)_3C\ CH_2OH$

Fig. 6. Hydrogen-suppressed graphs of aliphatic alcohols illustrating the equivalent treatment of first-row atoms of carbon and oxygen.

leads to a significant improvement in the correlation with boiling point and water solubility for 51 alcohols. The comparison for a few of these alcohols is shown in Table XI. Some examples of this empirical approach are illustrated in Fig. 7.

One shortcoming of this approach is that it does not appear to lend itself to easily calculated topological indices for molecules containing more than one heteroatom or different kinds of heteroatoms.

2. Nonempirical Treatment

An alternative and perhaps more fundamental approach is to reconsider the valence assigned to the heteroatom in the graph. To distinguish between an oxygen and a carbon atom in diethyl ether, it is necessary to distinguish between the valencies of the two, in spite of their identical formal connectivity in the hydrogen-suppressed graph. The concept of the rooted graph may be used as a basis for this approach.

One approach to this problem has been reported [12]. The objective is to develop valence values for heteroatoms that are nonempirical in the

TABLE XI

Alcohol Boiling Point Using an Empirical Scheme to Destroy Redundancies

Alcohol	$^1\chi$	Boiling point		
		Calc[a]	Calc[b]	Obs
2-Pentanol	2.769	119.9	119.2	119.0
2-Methylbutanol	2.769	119.9	131.7	128.7
3-Methylbutanol	2.808	121.5	130.3	131.2
3-Pentanol	2.808	121.5	120.6	115.3
3-Hexanol	3.307	141.0	139.9	135.4
3-Methylpentanol	3.307	141.0	126.2	122.4

[a] Calculated from Eq. (5) in Hall *et al.* [6] using only $^1\chi$. Note that the coefficient of $^1\chi$ in Eq. (5) should be +39.13.

[b] Calculated from Eq. (A-3) in Hall *et al.* [6]. b.p. = $-36.23 + 38.41\ ^1\chi_{CH} + 123.5\ c_{OH}$. $^1\chi_{CH}$ is $^1\chi$ for the hydrocarbon portion of the graph: $^1\chi_{CH} = {}^1\chi - c_{OH}$.

sense of formal connectivity and consonant with the underlying electronic structures of heteroatoms in molecules. The assignment of the valencies derived from a consideration of the origin of valencies of carbon atoms in an alkane. The δ values can be considered to be the difference between the number of valence electrons Z^v and the number of hydrogen atoms h_i (which are suppressed in the graph):

$$\delta_i^v = Z^v - h_i$$

Thus a methyl group δ is $4 - 3 = 1$, a methylene $\delta = 4 - 2 = 2$, and a methine $\delta = 4 - 1 = 3$. A quarternary carbon has a $\delta = 4 - 0 = 4$.

Oxygen Atoms. Applying this formulation to oxygen in an alcohol, $\delta^v = Z^v - h_i = 6 - 1 = 5$. The rooted graphs for the alcohols in Fig. 7 are shown in Fig. 8 along with the computed $^1\chi$ values, designated as $^1\chi^v$ to reflect the use of the valence δ. The redundancies inherent in Fig. 7 are destroyed by using the new δ value for the oxygen in a hydroxyl group. Correlation with alcohol boiling point based on this δ for oxygen is revealed in Table XII [12].

The differentiation between classes of alcohols arises from the different c_k values computed for each class of C—OH bond. Primary alcohols contain a $(10)^{-1/2} = (0.316)$ term, secondary alcohols a $(15)^{-1/2} = (0.258)$

term, and tertiary alcohols a $(20)^{-1/2} = (0.224)$ term. These values are clearly different from primary, secondary, and tertiary methyl groups (0.707, 0.577, 0.500, respectively), thereby removing the redundancies found in $^1\chi$.

In simple connectivity terms, the valence of 5 for a hydroxyl oxygen atom may also be thought of as arising from a consideration of the lone pair electrons on the atom. A hypothetical bond can be assumed to be formed with each lone pair electron. This bond would be formed with

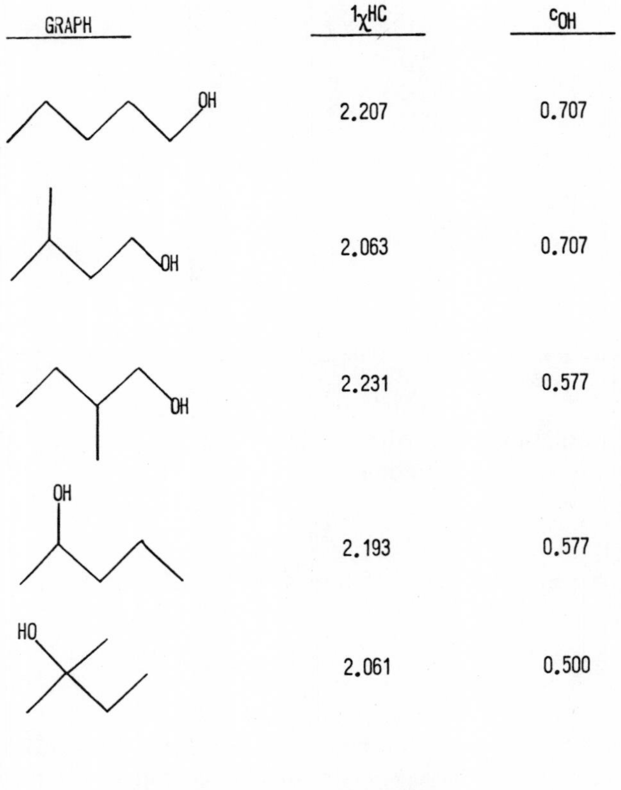

GRAPH	$^1\chi^{HC}$	c_{OH}
	2.207	0.707
	2.063	0.707
	2.231	0.577
	2.193	0.577
	2.061	0.500

$$\text{PROPERTY} = A_0 + A_1 \; ^1\chi^{HC} + A_2 \; c_{OH}$$

Fig. 7. Illustration of the empirical approach to heteroatoms with aliphatic alcohols. The $^1\chi$ value for the hydrocarbon portion of the graph $^1\chi^{HC}$, along with the edge connectivity for the C—OH bond c_{OH}, are entered into the relation between a property and connectivity in the equation.

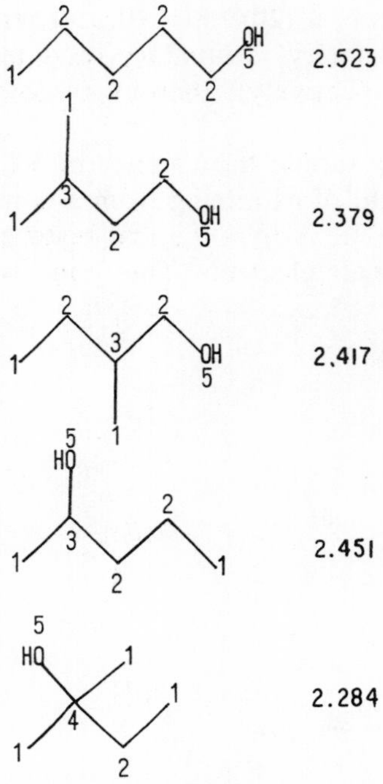

Fig. 8. Rooted graphs for alcohols showing the valence delta value for the OH group and the computed $^1\chi^v$ values.

another hypothetical atom with a valence of zero. The graph for ethanol, using the new valence assignments is

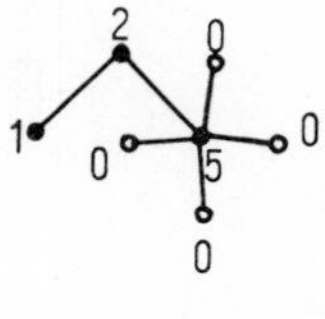

$(\underline{V})$

The edges or c_k terms composing χ^v are thus $(1,2) + (2,5) + 4\ (0,5)$. The four $(0,5)$ terms have $c_k = 0$ and hence, do not contribute to the value of χ^v.

An alternative derivation of the hydroxyl oxygen valence comes from the loop concept of connectivity. Each lone pair of electrons can be considered to form a loop (VI). Each loop increases the number of connections to the oxygen by 2 while not adding another edge term.

(VI)

The scheme $\delta_i^v = Z^v - h_i$ has inherent in it the ability uniquely to describe the valence of a carbonyl oxygen as distinguished from an hydroxyl oxygen. For a carbonyl oxygen atom $\delta^v = Z^v - h = 6 - 0 = 6$. The valence of six may be justified by either of the two previous derivations. Using acetone as an example (VII), the two-edge assignment for the double-bond connection will be discussed later.

(VII)

It follows from this formulation scheme that a protonated alcohol is represented by an oxygen valence $\delta^v = Z^v - h = 6 - 2 = 4$.

Nitrogen Atoms.   This same nonempirical derivation of heteroatom graph valencies can now be applied to another first-row atom, nitrogen. Consider the several classes of amine nitrogen atoms. Using $\delta^v = Z^v - h$, the valencies of each class can be written

$$CH_3—NH_2: \quad \delta_N^v = Z^v - h = 5 - 2 = 3$$

$$(CH_3)_2NH: \quad \delta_N^v = Z^v - h = 5 - 1 = 4$$

$$(CH_3)_3N: \quad \delta_N^v = Z^v - h = 5 - 0 = 5$$

The edge terms c_k for each case are 0.577, 0.500, and 0.447, respectively.

Nitrogen atoms can occur in other molecular situations. Using the same formula, the nitrogen valence in an alkylammonium ion, $R—NH_3^+$, would be $\delta^v = 5 - 3 = 2$. Tetra alkyl ammonium ions are considered to be tertiary nitrogen atoms to which an additional edge is connected; thus, $\delta^v = 5 - 0 + 1 = 6$. A nitro group nitrogen atom is considered to be formed from a nitroso group ($\delta^v = Z^v - h = 5 - 0 = 5$) by the addition of an edge to give $\delta^v = 5 - 0 + 1 = 6$. An *N*-oxide nitrogen atom also has a $\delta^v = 6$. Finally, the ammonium ion has $\delta^v = Z^v - h = 5$

TABLE XII

Correlation of Alcohol Boiling Point with $^1\chi^v$. Calculation Based on the Valence Delta, δ^v

		Boiling point (°C)	
Compound		Obs	Calc
1.	n-Butanol	117.7	114.56
2.	2-Methylpropanol	107.9	108.74
3.	2-Butanol	99.5	97.40
4.	n-Pentanol	137.8	134.27
5.	3-Methylbutanol	131.2	128.25
6.	2-Methylbutanol	128.7	129.74
7.	2-Pentanol	119.0	116.91
8.	3-Pentanol	115.3	118.39
9.	3-Methyl-2-butanol	111.5	111.76
10.	2-Methyl-2-butanol	102.0	102.33
11.	n-Hexanol	157.0	153.58
12.	2-Hexanol	139.9	136.42
13.	3-Hexanol	135.4	137.90
14.	3-Methyl-3-pentanol	122.4	124.23
15.	2-Methyl-2-pentanol·	121.4	121.85
16.	2-Methyl-3-pentanol	126.5	132.75
17.	3-Methyl-2-pentanol	134.2	132.75
18.	2,3-Dimethyl-2-butanol	118.6	117.08
19.	3,3-Dimethylbutanol	143.0	139.81
20.	3,3-Dimethyl-2-butanol	120.0	123.86
21.	4-Methylpentanol	151.8	147.77
22.	4-Methyl-2-pentanol	131.7	130.61
23.	2-Ethylbutanol	146.5	150.73
24.	n-Heptanol	176.3	173.09
25.	2-Methyl-2-hexanol	142.5	141.36
26.	3-Methyl-3-hexanol	142.4	143.74
27.	3-Ethyl-3-pentanol	142.5	146.12
28.	2,3-Dimethyl-2-pentanol	139.7	130.99
29.	2,3-Dimethyl-3-pentanol	139.0	138.97
30.	2,4-Dimethyl-2-pentanol	133.0	135.54

$- 4 = 1$, in a unique case. The valencies derived from $\delta^v = Z^v - h$ for nitrogen and oxygen in all situations are listed in Table XIII.

Redundant values of $^1\chi$ are possible in a series of primary amines. In Table XIV are listed several primary amines, the boiling points, and $^1\chi$ and $^1\chi^v$ values. Clearly, the use of $^1\chi^v$ destroys redundancies and significantly improves the relationship of the connectivity index to the physical property.

The redundancies found in $^1\chi$ for alcohols and amines previously described are due to the failure to distinguish between a carbon atom and the heteroatom in the graph. By selecting the delta value for the heteroatom in the graph different from the carbon atom, a differentiation is achieved and the redundancy destroyed. In principle, any value for the δ of the heteroatom, other than the δ for a carbon atom with the same degree of branching, would be effective. In this procedure, a prescription for heteroatom δ^v values, the objective has been to stay within the principles of graph theory and to find nonempirical values.

A second goal in the development of a set of heteroatom valencies is the creation of δ^v values that would permit the topological description of molecules containing both heteroatoms in a list or even both heteroatoms in the same molecule. Thus, a set of δ^v values could become self-consistent in their ability to relate to some physical properties.

A test of the internal consistency of the set of values in Table XIII can be made using molar refraction as a physical property [12]. For a

TABLE XIII

Heteroatom Valence Delta Values

Group	δ^v	Group	δ^v
NH_4^+	1	H_3O^+	3
NH_3	2	H_2O	4
—NH_2	3	—OH	5
—NH—	4	—O—	6
=NH	4	=O	6
—N—	5	O (both nitro)	6
≡N	5	O (both carboxylate)	6
=N— (pyridine)	5	—F	(−)20
—N— (quaternary)	6	—Cl	0.690
—N≡ (nitro)	6	—Br	0.254
—S—	0.944	—I	0.085
=S=	3.58		

TABLE XIV

Boiling Point of Primary Amines Predicted from $^1\chi$ and from Both $^1\chi$ and $^1\chi^v$ Values

Amine	Boiling point (obs)	Predicted From $^1\chi$	Predicted From $^1\chi$ and $^1\chi^v$
1. *tert*-Butylamine	46	52.3	46.8
2. *n*-Propylamine	49	47.8	52.0
3. *sec*-Butylamine[a]	63	66.3	64.1
4. Isobutylamine[a]	69	66.3	70.1
5. *n*-Butyl	77	73.8	77.4
6. 2-Methyl-*sec*-butylamine	78	81.5	75.4
7. 3-Pentylamine[c]	91	94.4	91.4
8. *sec*-Amylamine[b]	92	92.4	89.5
9. Isoamylamine[b]	96	92.4	95.5
10. 2-Methylpentylamine[c]	96	94.4	97.4
11. *n*-Pentylamine	104	99.9	102.8
12. Isohexylamine[d]	125	118.5	120.9
13. *n*-Hexyl	130	126.0	128.2
14. 2-Hexylamine[d]	114.5	118.5	114.9
15. 4-Heptylamine	139.5	146.5	142.3
16. *n*-Heptylamine	155	152.1	153.6
17. *n*-Octyl	180	178.1	179.0
18. *n*-Nonylamine	201	204.2	204.4
Correlation coefficient		0.996	0.999
Standard deviation		4.15	2.20

[a,b,c,d] These pairs of compounds represent cases with redundant $^1\chi$ values.

representative selection of molecules, containing nitrogen, oxygen, or both atoms, the $^1\chi^v$ values along with the $^1\chi$ terms will be shown in subsequent chapters. A list of δ^v values for nitrogen and oxygen in various bonding systems is found in Table XIII.

Halogen Atoms. A versatile connectivity index must be applicable to most atoms of interest to the chemist. This certainly includes the halogens. Except for fluorine, the halogens are in quantum levels higher than carbon. The topological significance of this fact is not clear at this time. The initial approach to fluorine connectivity is the formulation just developed.

Using the expression $\delta^v = Z^v - h_i$, the δ^v value for fluorine is 7. Testing this value for fluorobenzene, however, shows that the molecule is not adequately described by this valence. The predicted value of the molar refraction is 16.53 compared to the observed value of 26.03. It is apparent that the fluorine contributes about as much as a hydrogen atom on a benzene in the case of molar refractivity.

The $\delta^v = Z^v - h$ formula yields δ^v values of 7 for each halogen atom; however, the contributions of each halogen increase with the atomic number. It is apparent that the electronic structure of each halogen is decisive in influencing many properties, in addition to connectivity.

One approach to the connectivity of halogens has been to develop empirical values of valencies [12]. The molar refraction values in Table XV were used to calibrate the halogen valencies [12]. The objective was to create δ^v values consistent with carbon, nitrogen, and oxygen. A list of these δ^v values is found in Table XIII. It should be noted that the δ^v value for fluorine is a value leading to an edge term c_k that is negative, corresponding to the negative root of $(\delta_i\delta_j)^{-1/2}$. As a result, this contribution to the $^1\chi^v$ value is subtracted. As an example, the computation of $^1\chi^v$ for hexylfluoride is (VIII)

$$^1\chi^v = (1,2) + 4(2,2) - (2,20) = 0.707 + 2.000 - 0.158 = 2.549$$

(VIII)

These values were further tested against boiling points for a variety of monohaloaklanes (Table XVI). The results reveal a consistency among the halogen molecules.

Sulfur.   Sulfur-containing molecules were treated empirically, like the halogens, to arrive at δ^v values for the computation of $^1\chi^v$ [12]. These values, listed in Table XIII, were derived from molar refraction data in Table XV. It is noteworthy that the δ_S^v value developed for —SO_2— CH_3 permits a close prediction of the molar refraction of a molecule containing the —SO_2NH_2 group.

B. Unsaturated Bonds

The graph-theoretical consideration of an unsaturated molecule has been independently suggested by Randić [13] and Murray *et al.* [7]. The basic assumption is that a multiple bond in a molecule should be represented by endpoint valencies derived from the formula $\delta_i^v = Z^v - h_i$. This gives a $^1\chi^v$ value different from a $^1\chi$ value.

TABLE XV

Molar Refraction of Benzene Substituents and Values Predicted from Connectivity

Benzene substituent	$R_m{}^b$	R_m predicted from $^1\chi$ and $^1\chi^v$
1. CH_3-	4.7	4.1
2. CH_3CH_2-	9.4	9.6
3. $CH_3CH_2CH_2-$	14.0	14.5
4. $(CH_3)_2CH-$	14.0	13.3
5. $CH_3CH_2CH_2CH_2-$	18.7	19.4
6. $(CH_3)_3C-$	18.5	16.3
7. C_6H_5-	24.3	22.8
8. $-OH$	1.5	2.2
9. $-OCH_3$	6.5	6.4
10. $-OCH_2CH_3$	11.3	11.9
11. $-OCH_2CH_2CH_3$	15.9	16.8
12. $-OCH(CH_3)_2$	16.0	15.7
13. $-OCH_2CH_2CH_2CH_3$	20.7	21.7
14. $-OCH_2CH_2CH_2CH_2CH_3$	25.3	26.6
15. $-OC_6H_5$	26.6	25.2
16. $-O-CO-CH_3$	11.6	12.2
17. $-NH_2$	4.2	3.0
18. $-NH-CO-CH_3$	14.6	12.9
19. $-NO_2$	6.0	6.8
20. $-CHO$	5.3	5.7
21. $-CO-CH_3$	9.9	9.8
22. $-CO-O-CH_3$	11.4	12.0
23. $-CO-O-CH_2CH_3$	16.2	17.6
24. $-CO-NH_2$	8.8	8.6
25. $-CN$	5.2	5.9
26. $-F$	-0.4	a
27. $-Cl$	4.8	a
28. $-Br$	7.6	a
29. $-I$	12.8	a
30. $-SCH_3$	13.0	a
31. $-SO_2CH_3$	12.5	a
32. $-SO_2NH_2$	11.3	10.9
33. $-CF_3$	4.0	3.1
Correlation coefficient		0.990
Standard deviation		1.0

^{a} Observed values used to calculate the valence delta value for these cases. Hence, no calculated value is shown.

^{b} These values taken from Norrington *et al.* [14].

TABLE XVI

Boiling Point Prediction for Mixed Alkyl Halides Based on Valence Delta Values

				Boiling point	
Compound		$^1\chi$	$^1\chi^v$	Obs	Calc
1.	Ethylchloride	1.414	1.558	13	17.9
2.	Ethylbromide	1.414	2.110	38	38.9
3.	Ethyliodide	1.414	3.132	72	77.8
4.	Isopropylchloride	1.732	1.850	36.5	35.4
5.	Isopropylbromide	1.732	2.300	60	52.5
6.	Isopropyliodide	1.732	3.135	90	84.3
7.	*n*-Propylchloride	1.914	2.058	46.5	46.9
8.	*n*-Propylbromide	1.914	2.610	71	67.9
9.	*n*-Propyliodide	1.914	3.632	102.5	106.8
10.	2-Methyl-1-chloropropane	2.270	2.414	69	67.5
11.	2-Methyl-1-bromopropane	2.270	2.966	91	88.6
12.	2-Methyl-1-iodopropane	2.270	3.988	120	127.5
13.	*tert*-Butylchloride	2.000	2.102	51	50.3
14.	*tert*-Butylbromide	2.000	2.492	72.5	65.1
15.	*tert*-Butyliodide	2.000	3.215	98	92.7
16.	*n*-Butylchloride	2.414	2.558	78	75.9
17.	*n*-Butylbromide	2.414	3.110	101	96.9
18.	*n*-Butyliodide	2.414	4.132	130	135.8
19.	2-Chloropentane	2.770	2.888	97	95.5
20.	2-Bromopentane	2.770	3.338	117	112.7
21.	2-Iodopentane	2.770	4.173	142	144.5
22.	*n*-Propylfluoride	1.914	1.049	2.5	8.5
23.	*n*-Butylfluoride	2.414	1.549	32.5	37.5
24.	*n*-Pentylfluoride	2.914	2.049	62.8	66.5
Correlation coefficient				0.992	
Standard deviation				4.8	

The question then arises as to whether the multiple bonds should be counted as one edge [7] or two edges [11] in the graph. Randić bases his argument on the fact that a double-edge count is a more accurate reflection of the double bond in contrast to a single bond, connecting vertices with the same valence. An example is provided by the penta-1,3-diene molecule (IX). The graph and computation of $^1\chi^v$ according to

(IX)

Murray *et al.* [7] would be as shown in X. Bonds b and c give the same c_k terms in computing $^1\chi^v$, although bond b is a double bond while bond c is a single bond.

(X)

The computation of $^1\chi^v$ according to Randić is

$$^1\chi^v = (1,3) + 2(3,3) + (3,3) + 2(2,3) = 2.394$$

The $^1\chi^v$ value is higher as a result of the counting of each bond in the double bond in summing the c_k values. The graph thus has double edges for connections b and d, which differentiate these from edge c:

(XI)

A preliminary analysis of these two treatments does not reveal any major difference in ordering $^1\chi^v$ values for a set of alkenes. A comparison of both $^1\chi^v$ values with alkene boiling points shows a better correlation with $^1\chi^v$ computed by Murray's approach. When $^1\chi$ and either $^1\chi^v$ are multiply regressed against boiling point, both pairs of indices improve the correlation, with the $^1\chi^v$ computed according to Randić being very slightly better.

At this juncture, it would be desirable to retain the option of using either scheme to compute $^1\chi^v$ for double bonds.

The treatment of multiple bonds extends to triple bonds and heteroatom multiple bonds such as $C{=}O$, $C{\equiv}N$, and $X{=}O$. The counting of each bond once in the graph of molecules studied in Table XV was the procedure employed to correlate with molar refraction and to calibrate the empirical valence values of the halogens and sulfur in Table XIII.

C. Cyclic Molecules

The graph representations of pentane and cyclopentane reveal fundamental topological differences in these two types of molecules. The

graph for pentane (XII) has four edges, while the graph of cyclopentane
(XIII) has five. The computed connectivity indices are 2.414 and 2.500,

$^1\chi = 2.500$ $^1\chi = 2.414$

(XII)   (XIII)

respectively. These indices are in the correct relative order of the boiling
points (pentane, 36.1°; cyclopentane, 49.3°), and the refractive indices
(pentane, 1.3577; cyclopentane, 1.4068) but not for water solubility,
$-\log S$ (pentane, 3.27; cyclopentane, 2.65). For density (pentane,
0.6262; cyclopentane, 0.7454) the order is correct but the magnitude for
the cyclic is out of place in a list of straight-chain and cyclic alkanes. It
appears that for some properties the counting of the additional edge
gives an exalted value for the connectivity index mitigating against a
successful correlation.

We have approached this problem from the point of view of the graph-
theoretical concept of spanning trees. We can consider this problem with
the illustration of 1,3-dimethylcyclohexane as shown in Fig. 9. The

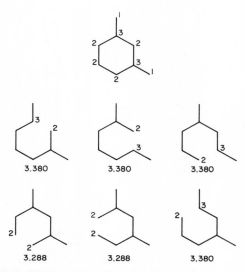

Fig. 9. Spanning-tree modification for cyclics showing all spanning trees, retaining original
graph valencies for each edge removal.

computed χ value is 3.788. The range of values for n-octane to dimethyl substituted hexanes is 3.914–3.664. For some properties such as solubility this 8-edge cyclohexane behaves as if it possessed only 7 edges.

The spanning trees of the substituted cyclohexane are also shown in Fig. 9. The vertex valencies of the original graph are appended to the vertices from which the edge is removed. Below each spanning tree is the value computed for χ with one less term, corresponding to the missing edge. Thus, the average difference between χ for the original graph and that for all the spanning trees is −0.439. For a nonsubstituted cyclohexane the corresponding difference is −0.5 and the value varies slightly for various patterns of disubstitution.

As a first approximation [5] we have used 0.5 as a modifying term to be subtracted from χ for a monocyclic irrespective of the degree or

TABLE XVII

Solubility of Cycloalkanes Predicted with Ring Modification Term

Compound	—log S	$^1\chi^a$	$^1\chi_{RC}^b$	$^1\chi_{STC}^c$
Cyclohexane	3.18	3.000	2.500	2.500
Methylcyclohexane	3.85	3.394	2.894	2.883
Cycloheptane	3.52	3.500	3.000	3.000
1-*cis*-2-Dimethylcyclohexane	4.26	3.805	3.305	3.321
Cyclooctane	4.15	4.000	3.500	3.500
Cyclopentane	2.65	2.500	2.000	2.000
n-Butane	2.63	1.914		
Isobutane	2.55	1.732		
n-Pentane	3.27	2.414		
Isopentane	3.18	2.270		
2,2-Dimethylpropane	3.13	2.000		
n-Hexane	3.95	2.914		
3-Methylpentane	3.83	2.808		
2,2-Dimethylbutane	3.67	2.561		
n-Heptane	4.53	3.414		
2,4-Dimethylheptane	4.39	3.126		
n-Octane	5.24	3.914		
2,2,4-Trimethylpentane	4.13	3.417		
Correlation coefficient		0.840	0.9537	0.9536
Standard deviation		0.61	0.225	0.225

[a] Branching index of Randić.

[b] Branching index with constant ring modification term of Kier *et al.* [5].

[c] Branching index with average spanning tree modification; see text.

pattern of substitution. It may well be that a weighted average, dependent on the pattern of substitution, should be used.

The retention of the original vertex valencies in writing the spanning trees of Fig. 9 is arbitrary. It may be that the actual subgraph valencies in each spanning tree should be used. However, such a procedure leads to a corresponding modifying term with a miniscule value of 0.041. Hence, retention of the valencies in G_v when dealing with spanning trees appears to be a more reasonable procedure.

The application of the cyclic modification term is illustrated with hydrocarbon solubility data as given by Amidon (Table XVII). The term $^1\chi$ refers to the branching index of Randić; $^1\chi_{RC}$ is obtained from $^1\chi$ by the subtraction of a constant 0.5 for the presence of a ring. The $^1\chi_{STC}$ is obtained by taking the average $^1\chi$ for all the spanning trees of the graph.

We have found that the use of higher order $^m\chi_t$ terms, in a multiple regression, is a successful approach in studies on lists of molecules containing both cyclic and noncyclic entries. Thus the ring correction is not required when more than one $^m\chi_t$ term can be used.

Molecular connectivity, as we have developed it, is still in its embryonic stages. The tenets set out in this chapter should provide investigators with methods and parameters to conduct large numbers of useful studies in organic and medicinal chemistry. In the last chapter we will describe some fertile areas for future work.

Symbols

$^m\chi_t^v$	the general symbol for a term in the connectivity function:
m	the subgraph order, number of edges in the corresponding subgraph
v	indicates the use of the valence deltas δ^v in computing subgraph terms (Note: the absence of the superscript v indicates the use of the simple connectivity deltas δ.)
t	subgraph type designation: P, path; C, cluster; PC, path/cluster; CH, chain; when no type designation t is given, assume a path
δ_i	the simple vertex (atom) connectivity, the number of edges connected to an atom or vertex i
δ_i^v	the valence connectivity
Z^v	the number of valence electrons
h_i	the number of hydrogen atoms attached to atom i
$^m S_j$	the connectivity value calculated for subgraph j; the sum of these terms for all subgraphs gives $^m\chi_t$
c_k	the edge connectivity term for $^1\chi$; $c_k = (\delta_i \delta_j)^{-1/2}$ for the kth edge
$^m\chi_n$	the chi term for an unbranched graph (e.g., as in a normal alkane)
$\Delta\chi_i$	the degree of branching defined as $\Delta\chi_i = {}^1\chi_n - {}^1\chi_i$
n	the number of vertices in the graph
N	the number of observations in a data set used in a regression analysis
r	the correlation coefficient
s	the standard deviation

References

1. H. Wiener, *J. Am. Chem. Soc.* **73**, 2636 (1947).
2. H. Hosoya, *Bull. Chem. Soc. Japan* **44**, 2332 (1971).
3. M. Randić, *J. Am. Chem. Soc.* **97**, 6609 (1975).
4. G. L. Amidon and S. T. Anik, *J. Pharm. Sci.* **65**, 801 (1976).
5. L. B. Kier, L. H. Hall, W. J. Murray, and M. Randić, *J. Pharm. Sci.* **64**, 1971 (1975).
6. L. H. Hall, L. B. Kier, and W. J. Murray, *J. Pharm. Sci.* **64**, 1974 (1975).
7. W. J. Murray, L. B. Kier, and L. H. Hall, *J. Pharm. Sci.* **64**, 1978 (1975).
8. L. B. Kier, L. H. Hall, W. J. Murray, and M. Randić, *J. Pharm. Sci.* **65** (1976).
9. L. B. Kier, W. J. Murray, and L. H. Hall, *J. Med. Chem.* **18**, 1272 (1975).
10. W. J. Murray, L. B. Kier, and L. H. Hall, *J. Med. Chem.* **19**, 573 (1976).
11. L. H. Hall, L. B. Kier, and W. J. Murray, unpublished work.
12. L. B. Kier and L. H. Hall, *J. Pharm. Sci.* **65** (1976).
13. M. Randić, manuscript in preparation.
14. F. E. Norrington, R. M. Hyde, S. G. Williams, and R. Wooten, *J. Med. Chem.* **18**, 604 (1975).

Chapter Four

MOLECULAR PROPERTIES AND CONNECTIVITY

In a previous chapter we used the terms additive and constitutive in discussing properties of organic molecules. Another classification categorized properties on the basis of their dependence on state of aggregation. A property that is independent of state and is based on the intrinsic nature of the isolated molecule is said to be a *molecular property*. Other properties depend on the state of aggregation. Such properties that arise from the forces at play in condensed phases are called *molar* or *bulk properties*. Many commonly encountered properties depend on the nature of isolated molecules as well as interactions between molecules. Hence, examples of purely molecular or molar properties are limited.

For convenience, we divide our study of physical and chemical properties into the molecular and molar categories. This is a useful classification because the chemist or biologist has some intuitive feeling for the nature of properties divided in this manner.

The biological or organic researcher is familiar with bond energy terms derived from heat of atomization, with molar refraction and its relation to molecular polarizability and volume, as well as with liquid density and its relation to molar volume. Heat of atomization is a prime example of a molecular property. Liquid density falls into the molar category. Molar refraction R_m is derived from refractive index n and density d measurements:

$$R_m = \frac{n^2 - 1}{n^2 + 2} \frac{M}{d}$$

Refractive index depends on state of aggregation and, hence, is a molar property. However, molar refraction is independent of state and is a measure of molecular volume. Molar refraction is thus a molecular property.

The central question in this book is whether results obtained from molecular connectivity calculations bear a significant relationship to physicochemical and biological properties of molecules. In the next two chapters we will examine physicochemical properties for which molecular connectivity demonstrates some promise in revealing structure–activity relationships. We will examine these properties, grouped largely as molecular or bulk dependent.

I. Heat of Atomization and Formation

The nature of molecular connectivity may be studied by examining the relationships between structure and certain thermochemical quantities. Further, reliable experimental data are available for a wide variety of organic molecules [1].

Let us consider what property might be most meaningfully related to structure. The total energy content of an isolated molecule depends on the intramolecular energy from the chemical binding of the atoms as well as translational, rotational, and vibrational energy. However, it is convenient to think of the total energy of a molecule in terms of the following process:

$$\text{molecule (ground state, ideal gas, } T)$$

$$\rightarrow \text{atoms (ground state, ideal gas, } T)$$

The energy for this process is known as the energy of atomization. A more convenient quantity, the *heat of atomization* ΔH_a, is the energy of atomization with the $\Delta(PV)$ term in the well-known thermodynamic relation between enthalpy H and internal energy E,

$$\Delta H_a = \Delta E_a + \Delta(PV)$$

We may equate the heat of atomization to the total chemical binding energy of the molecule, which is composed of three terms:

1. The sum of the chemical bond energies and bond interactions.
2. Destabilizing effects ascribed to steric repulsion and strain.
3. Additional stabilization arising from electron delocalization effects.

The combination of these effects may be written in general fashion as

$$\Delta H_a = \Sigma \text{ chemical bond energies} - \Sigma \text{ destabilization energies}$$

$$+ \Sigma \text{ stabilization energies}$$

Such dissection of total chemical binding energy ΔH_a may be the basis of a study of the topological nature of this quantity. Our later discussion will examine the question further.

The heat of formation ΔH_f is defined as the enthalpy change for the following chemical process:

elements (standard states) → compound (standard state)

For example, ΔH_f^{gas} for ethanol is obtained from the measured heat at constant pressure Q_P for the following process:

$$2C \text{ (graphite)} + 3H_2(g) + \tfrac{1}{2}O_2(g) = C_2H_5OH(g), \qquad \Delta H_a = Q_P$$

The heat of formation in the gas phase is directly related to the heat of atomization of the compound and of the constituent elements as shown by the following equation, expressed for a general compound $C_nH_mO_l$:

$$\Delta H_a = n \, \Delta H_f^\circ(C_{gas}) + m \, \Delta H_f^\circ(H_{gas}) + l \, \Delta H_f^\circ(O_{gas}) - \Delta H_f^\circ(C_nH_mO_{l \text{ gas}})$$

Since molecular connectivity calculations are made on individual molecules, a critical test of the ability of this method is the quality of the relationship of the connectivity function to molecular properties such as heat of atomization and heat of formation.

A. Heat of Atomization for Hydrocarbons

The heat of atomization can be shown to bear a significant relationship to the number of first-row atoms n. The regression coefficients and statistical data are recorded in Table I. It appears that n is required as a regression variable for heat of atomization because ΔH_a is linearly dependent on n as well as the basic experimental quantity ΔH_f^{gas} for hydrocarbons:

$$\Delta H_a = n \, \Delta H_a \text{ (graphite)} + (2n + 2) \, \Delta H_a \text{ (H atoms)} - \Delta H_f^{gas}$$

By testing a number of terms in the connectivity function, a significant improvement in the predicted values can be achieved [2]. When $^1\chi$ is added as a regression variable, the standard error decreases 35% to 0.96 as shown in Table I. As other subgraph terms are added, the correlation continues to improve. An eight-term expression yields a standard error of 0.37 kcal, near to the estimated experimental error, 0.32 kcal, as given by Zwolinski [3b]. Calculated values are given in Table II.

This correlation is a significant achievement and tends to establish the method of molecular connectivity as a fundamental approach to the correlation of physicochemical properties with structure. A seven- or eight-term connectivity function produces nearly the same quality fit as the seven-variable function of Zwolinski [3a]. The Zwolinski approach

TABLE I

Correlation of Alkane Heat of Atomization ΔH_a with the Connectivity Function

Statistical data			Regression coefficients							
r	s	n	$^1\chi$	$^4\chi_P$	$^5\chi_P$	$^5\chi_C$	$^4\chi_{PC}$	$^5\chi_{PC}$	$^6\chi_{PC}$	Constant
0.9999	1.455	280.53	−6.321							115.72
>0.9999	0.960	283.33								116.19
>0.9999	0.464	286.38	−12.46	1.515	—	−2.474	1.1420	−2.0260	—	114.38
>0.9999	0.371	286.15	−12.08	0.9209	1.5046	−2.443	0.8595	−0.4996	−1.421	114.65

TABLE II

Observed and Calculated Heat of Atomization ΔH_a for Alkanes from Correlation with the Connectivity Function

		ΔH_a	
Compound		Obs	Calc
1.	Propane	955.49	956.01
2.	n-Butane	1236.31	1236.12
3.	2-Methylpropane	1238.31	1238.32
4.	n-Pentane	1516.65	1516.55
5.	2-Methylbutane	1518.57	1518.32
6.	2,2-Dimethylpropane	1521.32	1521.23
7.	n-Hexane	1797.10	1797.17
8.	2-Methylpentane	1798.80	1798.71
9.	3-Methylpentane	1798.16	1798.23
10.	2,2-Dimethylbutane	1801.49	1801.34
11.	2,3-Dimethylbutane	1799.63	1799.95
12.	n-Heptane	2077.52	2077.59
13.	2-Methylhexane	2079.23	2079.22
14.	3-Methylhexane	2078.60	2078.60
15.	3-Ethylpentane	2077.97	2077.99
16.	2,2-Dimethylpentane	2081.91	2081.19
17.	2,3-Dimethylpentane	2080.26	2079.53
18.	2,4-Dimethylpentane	2080.92	2080.39
19.	3,3-Dimethylpentane	2080.81	2080.82
20.	2,2,3-Trimethylbutane	2081.59	2081.96
21.	n-Octane	2357.94	2358.05
22.	2-Methylheptane	2359.62	2359.62
23.	3-Methylheptane	2358.94	2359.11
24.	4-Methylheptane	2358.81	2358.97
25.	3-Ethylhexane	2358.52	2358.54
26.	2,2-Dimethylhexane	2361.83	2361.52
27.	2,3-Dimethylhexane	2359.25	2359.78
28.	2,4-Dimethylhexane	2360.56	2360.16
29.	2,5-Dimethylhexane	2361.33	2361.03
30.	3,3-Dimethylhexane	2360.73	2360.69
31.	3,4-Dimethylhexane	2359.03	2359.07
32.	2-Methyl-3-ethylpentane	2358.60	2359.01
33.	3-Methyl-3-ethylpentane	2359.50	2359.88
34.	2,2,3-Trimethylpentane	2360.73	2360.80
35.	2,2,4-Trimethylpentane	2361.69	2361.92
36.	2,3,3-Trimethylpentane	2359.85	2360.71
37.	2,3,4-Trimethylpentane	2360.09	2360.10
38.	2,2,3,3-Tetramethylbutane	2362.11	2362.19
39.	n-Nonane	2638.35	2638.51
40.	3,3-Diethylpentane	2639.05	2638.55
41.	2,2,3,3-Tetramethylpentane	2640.31	2639.92
42.	2,2,3,4-Tetramethylpentane	2640.25	2639.83
43.	2,2,4,4-Tetramethylpentane	2641.44	2642.09
44.	2,3,3,4-Tetramethylpentane	2640.07	2640.46

TABLE III

Correlation of Alkane Heat of Formation ΔH_f with the Connectivity Function

Statistical data			Regression coefficients								
r	s	n	$^1\chi$	$^2\chi$	$^3\chi_C$	$^4\chi_C$	$^5\chi_C$	$^4\chi_{PC}$	$^5\chi_{PC}$	$^6\chi_{PC}$	Constant
0.9959	1.32	6.087									12.24
0.9704	3.54	—	12.27								16.33
0.9971	1.13	7.649	−3.286								11.70
0.9995	0.384	—	10.09	−2.125	7.520	−11.78	−1.705	0.9186	−1.621		10.61
0.9996	0.376	—	10.15	−2.517	7.632	−12.02	−1.724	0.8915	−1.457	−0.2774	10.42

explicitly takes account of bond terms, bond interactions, and in addition, nonbonded steric terms. It is especially noteworthy that two of the three predicted values that deviate the most from the observed values, numbers 17 and 36, have been suggested by Zwolinski as candidates for reexamination of the experimental results.

B. Heat of Formation (Gas) and Connectivity for Hydrocarbons

The heat of formation in the gas phase can be demonstrated to bear a strong relationship to connectivity terms, as shown in Table III. The presence of n, the number of carbon atoms, is not required for a quality correlation.

A correlation with n and $^1\chi$ produces a standard error of about 1 kcal [2]. A more sophisticated treatment, using only seven or eight members in a connectivity function, reduces the standard error to about the level of the estimated experimental standard error, 0.32 kcal [3b]. Observed and calculated values for ΔH_f^{gas} are shown in Table IV based on line 5, Table III.

Once again the straightforward application of the connectivity function has led to a correlation within experimental error for highly accurate experimental data. Note also that no functions especially calibrated for molecular data have been used. Only sets of numerical structural descriptors based on the molecular graph are required. Agreement of such quality encourages the idea that molecular connectivity is a fundamental method of SAR.

C. Alcohol Heat of Atomization and Formation

In a first-order treatment of heteroatom-containing molecules the hydroxyl group of alcohols may be treated in a manner equivalent to methyl groups, to which they are isoelectronic [2]. Since $^1\chi$ decreases with branching while ΔH increases and since $^1\chi$ increases with the number of carbon atoms as does ΔH, the regression equation may be formulated as follows for both ΔH_a and ΔH_f^{gas} :

$$\Delta H_i = b_0 + b_1 \Delta\chi_i + b_2\chi_n$$

where χ_n is $^1\chi$ for the straight-chain alcohol of $n - 1$ carbons and $\Delta\chi_i = \chi_n - {}^1\chi_i$ where χ_i is $^1\chi$ for the ith alcohol of $n - 1$ carbons. Thus, $\Delta\chi_i$ is a measure of the degree of skeletal branching.

TABLE IV

Observed and Calculated Heat of Formation ΔH_f for Alkanes from Correlation with the Connectivity Function

		ΔH_f	
Compound		Obs	Calc
1.	Ethane	20.24	20.70
2.	Propane	24.82	24.88
3.	*n*-Butane	30.15	29.93
4.	2-Methylpropane	32.15	32.43
5.	*n*-Pentane	35.00	34.98
6.	2-Methylbutane	36.92	36.97
7.	2,2-Dimethylpropane	40.27	39.95
8.	*n*-Hexane	39.96	40.02
9.	2-Methylpentane	41.66	41.44
10.	3-Methylpentane	41.02	41.19
11.	2,2-Dimethylbutane	44.35	44.43
12.	2,3-Dimethylbutane	42.49	42.95
13.	*n*-Heptane	44.89	45.07
14.	2-Methylhexane	46.60	46.62
15.	3-Methylhexane	45.96	45.96
16.	3-Ethylpentane	45.34	45.49
17.	2,2-Dimethylpentane	49.29	48.14
18.	2,3-Dimethylpentane	46.65	46.62
19.	2,4-Dimethylpentane	48.30	47.21
20.	3,3-Dimethylpentane	48.17	48.19
21.	2,2,3-Trimethylbutane	48.96	49.31
22.	*n*-Octane	49.82	49.85
23.	2-Methylheptane	51.50	51.67
24.	3-Methylheptane	50.82	51.11
25.	4-Methylheptane	50.69	50.70
26.	3-Ethylhexane	50.40	50.44
27.	2,2-Dimethylhexane	53.71	53.55
28.	2,3-Dimethylhexane	51.13	51.67
29.	2,4-Dimethylhexane	52.44	52.01
30.	2,5-Dimethylhexane	53.21	53.34
31.	3,3-Dimethylhexane	52.61	52.39
32.	3,4-Dimethylhexane	50.91	50.73
33.	2-Methyl-3-ethylpentane	50.48	50.55
34.	3-Methyl-3-ethylpentane	51.38	51.59
35.	2,2,3-Trimethylpentane	52.61	52.44
36.	2,2,4-Trimethylpentane	53.57	53.18
37.	2,3,3-Trimethylpentane	51.73	52.52
38.	2,3,4-Trimethylpentane	51.97	51.46
39.	2,2,3,3-Tetramethylbutane	53.99	54.19
40.	*n*-Nonane	54.54	54.79

TABLE IV—Continued

Compound	ΔH_f	
	Obs	Calc
41. 4-Methyloctane	56.19	55.63
42. 2,2-Dimethylheptane	58.83	58.59
43. 2,2,3-Trimethylhexane	57.70	57.78
44. 2,2,4-Trimethylhexane	58.12	58.27
45. 2,2,5-Trimethylhexane	60.53	60.40
46. 2,3,3-Trimethylhexane	57.31	57.06
47. 2,3,5-Trimethylhexane	57.97	57.89
48. 2,4,4-Trimethylhexane	57.47	57.78
49. 3,3,4-Trimethylhexane	56.42	56.14
50. 2,2-Dimethyl-3-ethylpentane	55.37	55.90
51. 2,4-Dimethyl-3-ethylpentane	54.48	55.12
52. 3,3-Diethylpentane	55.81	54.98
53. 2,2,3,3-Tetramethylpentane	57.07	56.91
54. 2,2,3,4-Tetramethylpentane	56.81	56.69
55. 2,2,4,4-Tetramethylpentane	58.16	58.41
56. 2,3,3,4-Tetramethylpentane	56.68	56.28
57. *n*-Decane	59.67	59.67
58. 3,3,5-Trimethylheptane	62.22	62.60
59. 2,2,3,3-Tetramethylhexane	62.08	62.12
60. 2,2,5,5-Tetramethylhexane	67.29	67.24
61. *n*-Undecane	64.58	64.59
62. 2-Methyldecane	66.06	66.22
63. *n*-Dodecane	69.49	69.51
64. 2,2,4,4,6-Pentamethylheptane	72.10	72.54
65. 2,2,4,6,6-Pentamethylheptane	75.45	75.14
66. *n*-Hexadecane	89.21	89.20
67. 2-Methylpentadecane	90.87	90.83

The regression equations and statistical results are

$$\Delta H_a^i = -21.926 + 22.275 \, \Delta \chi_i + 559.975 \chi_n$$

$$r > 0.9999, \qquad s = 1.01, \qquad N = 19$$

$$\Delta H_f^{gas} = 41.518 + 21.510 \, \Delta \chi_i + 4.873 n$$

$$r = 0.9948, \qquad s = 1.04, \qquad N = 20$$

Observed and calculated values of ΔH_a and ΔH_f are shown in Tables V and VI. Either χ_n or n may be used as a regression variable [2] since

TABLE V

Correlation of Heat of Atomization ΔH_a with Connectivity for Aliphatic Alcohols

				ΔH_a	
Compound		$\Delta\chi$	n	Obs	Calc
1.	Methanol	0.00000	2.0	486.93	490.08
2.	Ethanol	0.00000	3.0	770.20	770.06
3.	*n*-Propanol	0.00000	4.0	1050.23	1050.03
4.	2-Propanol	0.18216	4.0	1054.18	1054.07
5.	*n*-Butanol	0.00000	5.0	1329.95	1330.01
6.	2-Methylpropanol	0.14416	5.0	1332.00	1333.20
7.	2-Butanol	0.14416	5.0	1334.14	1333.20
8.	2-Methyl-2-propanol	0.41421	5.0	1338.88	1339.18
9.	*n*-Pentanol	0.00000	6.0	1609.92	1609.99
10.	2-Pentanol	0.14416	6.0	1614.44	1613.18
11.	3-Pentanol	0.10616	6.0	1614.47	1612.34
12.	2-Methyl-1-butanol	0.14416	6.0	1611.45	1613.18
13.	3-Methyl-1-butanol	0.10615	6.0	1611.23	1612.34
14.	2-Methyl-2-butanol	0.35355	6.0	1618.33	1617.82
15.	3-Methyl-2-butanol	0.27148	6.0	1615.61	1616.00
16.	*n*-Hexanol	0.00000	7.0	1890.01	1889.97
17.	*n*-Heptanol	0.00000	8.0	2168.55	2169.94
18.	*n*-Octanol	0.00000	9.0	2449.86	2449.92
19.	2-Ethyl-1-hexanol	0.06816	9.0	2451.87	2451.43
20.	*n*-Nonanol	0.00000	10.0	2730.78	2729.90
21.	*n*-Decanol	0.00000	11.0	3009.57	3009.87

they are related as follows:

$$\chi_n = (n - 3)/2 + (2)^{1/2}, \qquad n > 2$$

It is observed from the experimental values that branching in the carbon skeleton increases ΔH about half as much as that attributed to the effect of hydroxyl group position. The $\Delta\chi$ can be partitioned between two terms representing these independent branching characteristics, $\Delta\chi^S$ and $\Delta\chi^{OH}$. The degree of branching $\Delta\chi$ is the same in both 2-methyl-1-butanol and 2-pentanol, 0.144. However, the branching in the former is attributed to skeletal branching; thus, $\Delta\chi^S = 0.144$ and $\Delta\chi^{OH} = 0.0$, whereas in the secondary alcohol the branching is due to hydroxyl group position, $\Delta\chi^S = 0.0$ and $\Delta\chi^{OH} = 0.144$. In both 2-methyl-2-butanol and 2-methyl-3-butanol the degree of branching is equally divided

TABLE VI

Correlation of Heat of Formation (Gas Phase) with Connectivity for Aliphatic Alcohols

Compounds		$\Delta\chi$	n	ΔH_f Obs	ΔH_f Calc^a	ΔH_f Calc^b
1.	Methanol	0.00000	2.0	48.07	51.26	51.21
2.	Ethanol	0.00000	3.0	56.24	56.13	56.10
3.	*n*-Propanol	0.00000	4.0	61.17	61.01	60.99
4.	2-Propanol	0.18216	4.0	65.12	64.92	64.91
5.	*n*-Butanol	0.00000	5.0	65.79	65.88	65.88
6.	2-Methylpropanol	0.14416	5.0	67.84	68.98	67.60
7.	2-Butanol	0.14416	5.0	69.98	68.98	70.37
8.	2-Methyl-2-propanol	0.41416	5.0	74.72	74.78	74.80
9.	*n*-Pentanol	0.00000	6.0	70.66	70.75	70.78
10.	2-Pentanol	0.14416	6.0	75.18	73.85	75.26
11.	3-Pentanol	0.10616	6.0	75.21	73.03	74.08
12.	2-Methyl-1-butanol	0.10616	6.0	72.19	73.03	72.04
13.	3-Methyl-1-butanol	0.14416	6.0	72.02	73.85	72.49
14.	2-Methyl-2-butanol	0.27148	6.0	75.35	76.58	76.61
15.	3-Methyl-2-butanol	0.35355	6.0	79.07	78.37	78.40
16.	*n*-Hexanol	0.00000	7.0	75.65	75.62	75.67
17.	*n*-Heptanol	0.00000	8.0	79.09	80.50	80.56
18.	*n*-Octanol	0.00000	9.0	85.30	85.37	85.46
19.	2-Ethyl-1-hexanol	0.06816	9.0	87.31	86.68	86.18
20.	*n*-Nonanol	0.00000	10.0	91.12	90.24	90.35
21.	*n*-Decanol	0.00000	11.0	94.81	95.12	95.24

^{a} Based on the two-variable equation using n and $\Delta\chi$. See text.
^{b} Based on the three-variable equation using n, $\Delta\chi^S$, and $\Delta\chi^{OH}$. See text.

between skeletal and functional group position effects. Thus, the $\Delta\chi^S$ and $\Delta\chi^{OH}$ variables have the same value as shown in Table VII. It is seen, then, that in a straightforward manner values may be deduced for the partitioned degree of branching variables.

There is significant decrease in the standard error, about 30–40%, when these variables are introduced into the regression equation [2]:

$$\Delta H_a^i = -69.965 + 12.389 \, \Delta\chi_i^S + 31.805 \, \Delta\chi_i^{OH} + 279.998n$$

$$r > 0.9999, \quad s = 0.622, \quad N = 20$$

$$\Delta H_f^{gas} = 41.424 + 11.905 \, \Delta\chi_i^S + 31.555 \, \Delta\chi_i^{OH} + 4.893n$$

$$r = 0.9976, \quad s = 0.719, \quad N = 20$$

TABLE VII

Heat of Atomization ΔH_a for Aliphatic Alcohols and Connectivity Based on Partitioning of Branching

					ΔH_a	
Compound	$\Delta\chi^S$	$\Delta\chi^{OH}$	n	Obs	Calc
1. Methanol	0.00000	0.00000	2.0	486.93	490.03
2. Ethanol	0.00000	0.00000	3.0	770.20	770.02
3. *n*-Propanol	0.00000	0.00000	4.0	1050.23	1050.02
4. 2-Propanol	0.09108	0.09108	4.0	1054.18	1054.04
5. *n*-Butanol	0.00000	0.00000	5.0	1329.95	1330.02
6. 2-Methylpropanol	0.14416	0.00000	5.0	1332.00	1331.81
7. 2-Butanol	0.00000	0.14416	5.0	1334.14	1334.60
8. 2-Methyl-2-propanol	0.20708	0.20700	5.0	1338.88	1339.17
9. *n*-Pentanol	0.00000	0.00000	6.0	1609.92	1610.02
10. 2-Pentanol	0.00000	0.14416	6.0	1614.44	1614.60
11. 3-Pentanol	0.00000	0.10616	6.0	1614.47	1613.39
12. 2-Methyl-1-butanol	0.14416	0.00000	6.0	1611.45	1611.80
13. 3-Methyl-1-butanol	0.10616	0.00000	6.0	1611.23	1611.33
14. 2-Methyl-2-butanol	0.17677	0.17677	6.0	1618.33	1617.83
15. 3-Methyl-2-butanol	0.13657	0.13657	6.0	1615.61	1616.05
16. *n*-Hexanol	0.00000	0.00000	7.0	1890.01	1890.02
17. *n*-Heptanol	0.00000	0.00000	8.0	2168.55	2170.01
18. *n*-Octanol	0.00000	0.00000	9.0	2449.86	2450.01
19. 2-Ethyl-1-hexanol	0.06816	0.00000	9.0	2451.87	2450.86
20. *n*-Nonanol	0.00000	0.00000	10.0	2730.78	2730.01
21. *n*-Decanol	0.00000	0.00000	11.0	3009.57	3010.01

Observed and calculated values are tabulated in Tables VI and VII along with values for $\Delta\chi^S$ and $\Delta\chi^{OH}$.

The partitioning of branching between effects attributed to skeletal branching or functional group position is an empirical approach to the heteroatom problem. Such a method is attractive because of its simplicity and its direct relation to intuitive notions of branching. Only one additional parameter is necessary and there is no need for tabulated heteroatom parameters.

The nonempirical method for heteroatoms introduced in Chapter 3 leads to a more general and sophisticated approach to the application of connectivity through the use of valence connectivity terms. The use of $^1\chi$ and $^1\chi^v$ together with χ_n produces a quality of fit essentially the same as with $\Delta\chi^S$ and $\Delta\chi^{OH}$ along with n. The standard error when valence terms are used is 0.64, as shown in Table VIII. The use of other subgraph terms produces an even lower standard error but the limited number of observations, 20, prevents an extensive investigation. The ΔH_a value for n-heptanol appears suspect since it deviates so much more than any other, as shown in Table IX. When n-heptanol is excluded, predicted values of other compounds are little affected and the standard error drops to 0.386 kcal, the probable level of experimental error. Only a four-variable connectivity function is required for this excellent correlation. For molecules containing heteroatoms in various positions of substitution and with various skeletal patterns, this result may be considered very good.

D.   Ethers and Thiols: Heat of Atomization

The first-order treatment, using nonhydrogenic atoms equivalently, has also been applied to ethers and thiols [2] with equal success as to hydrocarbons and alcohols. The regression equations and statistical data are:

Ethers

$$\Delta H_a^i = -83.031 + 13.033 \Delta\chi_i + 281.052n$$

$$r > 0.9999, \qquad s = 2.37, \qquad N = 12$$

TABLE VIII

Heat of Atomization of Aliphatic Alcohols and Connectivity Terms: Correlation and Statistical Data

Statistical data		Regression coefficients					
r	s	χ_n	$^1\chi$	$^1\chi^v$	$^2\chi^v$	$^3\chi_P^v$	Constant
0.9999	0.990	582.02	−22.07				−21.82
>0.9999	0.640	571.33	−48.42	37.11			−7.44
>0.9999	0.518	609.08	−41.97	—	−8.65	−1.91	−29.03
>0.9999[a]	0.386	608.97	−41.67	—	−8.71	−2.11	−29.24

[a] n-Heptanol excluded from regression because its residual is large (1.31) compared to the others.

TABLE IX

Observed and Predicted Heat of Atomization for Aliphatic Alcohols Correlated with the Connectivity Function

Compound	Heat of atomization ΔH_a		
	Obs	Calc[a]	Calc[b]
1. Ethanol	770.20	770.25	770.30
2. n-Propanol	1050.23	1049.84	1049.90
3. Isopropanol	1054.18	1054.24	1054.22
4. n-Butanol	1329.95	1329.80	1329.88
5. 2-Methylpropanol	1332.00	1331.82	1331.86
6. sec-Butanol	1334.14	1334.15	1334.16
7. tert-Butanol	1338.88	1338.71	1338.71
8. n-Pentanol	1609.92	1609.82	1609.93
9. 2-Pentanol	1614.44	1614.19	1614.26
10. 3-Pentanol	1614.47	1614.65	1614.74
11. 2-Methylbutanol	1611.45	1611.51	1611.52
12. 3-Methylbutanol	1611.23	1611.86	1611.92
13. 2-Methyl-2-butanol	1618.33	1618.11	1618.05
14. 2-Methyl-3-butanol	1615.61	1616.04	1615.99
15. n-Hexanol	1890.01	1889.84	1889.98
16. n-Heptanol	2168.55	2169.86	—
17. n-Octanol	2449.86	2449.88	2450.07
18. 2-Ethylhexanol	2451.87	2451.26	2451.36
19. n-Nonanol	2730.78	2729.90	2730.11
20. n-Decanol	3009.57	3009.92	3010.16

[a] Based on parameters from Table VIII, line 3.
[b] Based on parameters from Table VIII, line 4.

Thiols

$$\Delta H_a^i = -109.247 + 12.658\,\Delta\chi_i + 280.083n$$

$$r > 0.9999, \qquad s = 0.30, \qquad N = 14$$

Observed and predicted values are shown in Tables X and XI.

E.   Heat of Atomization for Monoolefins

For unsaturated hydrocarbons containing one double bond, molecular structure is composed of two independent features: branching in the skeleton and location of the double bond in the skeleton. Thus, in

TABLE X

Correlation of Heat of Atomization ΔH_a and Connectivity for Saturated Ethers

Compound		$\Delta \chi$	n	ΔH_a Obs	Calc
1.	Dimethyl ether	0.000000	3.0	757.95	760.12
2.	Methyl ethyl ether	0.000000	4.0	1040.78	1041.17
3.	Diethyl ether	0.000000	5.0	1324.42	1322.22
4.	Methyl-*n*-propyl ether	0.000000	5.0	1320.98	1322.22
5.	Methyl-*sec*-propyl ether	0.144159	5.0	1324.40	1324.10
6.	Methyl-*tert*-butyl ether	0.353554	6.0	1608.86	1607.88
7.	Di-*n*-propyl ether	0.000000	7.0	1884.11	1884.33
8.	Di-*sec*-propyl ether	0.288318	7.0	1890.46	1888.09
9.	Isopropyl-*tert*-butyl ether	0.497713	8.0	2174.86	2171.87
10.	Di-*n*-butyl ether	0.000000	9.0	2444.28	2446.43
11.	Di-*sec*-butyl ether	0.212308	9.0	2450.72	2449.20
12.	Di-*tert*-butyl ether	0.707107	9.0	2451.56	2455.65

TABLE XI

Correlation of Heat of Atomization ΔH_a with Connectivity for Alkyl Thiols

Compound		$\Delta \chi$	n	ΔH_a Obs	Calc
1.	Methanethiol	0.000000	2.0	450.35	450.91
2.	Ethanethiol	0.000000	3.0	731.05	731.00
3.	1-Propanethiol	0.000000	4.0	1011.29	1011.08
4.	2-Propanethiol	0.182163	4.0	1013.29	1013.39
5.	1-Butanethiol	0.000000	5.0	1291.23	1291.16
6.	2-Butanethiol	0.144159	5.0	1293.34	1292.99
7.	2-Methyl-1-propanethiol	0.144159	5.0	1293.42	1292.99
8.	2-Methyl-2-propanethiol	0.414214	5.0	1296.37	1296.41
9.	1-Pentanethiol	0.000000	6.0	1571.59	1571.25
10.	3-Methyl-1-butanethiol	0.144159	6.0	1572.76	1573.07
11.	2-Methyl-2-butanethiol	0.353554	6.0	1575.65	1575.72
12.	1-Hexanethiol	0.000000	7.0	1851.34	1851.33
13.	1-Heptanethiol	0.000000	8.0	2131.28	2131.41
14.	1-Decanethiol	0.000000	11.0	2971.50	2971.66

addition to the terms used in the equation for saturated hydrocarbons, two additional terms are used for χ^v and χ_n^v. These variables describe the connectivity associated with the double bond [2].

The regression equation and associated statistics are

$$\Delta H_a^i = 39.494 + 22.290 \, \Delta\chi_i + 221.228n - 11.615 \, \Delta\chi_i^v + 116.963\chi_n^v$$

$$r > 0.9999, \qquad s = 1.46, \qquad N = 37$$

Calculated values are shown in Table XII. It should be noted that the coefficient of the $\Delta\chi_i$ term is the same as in the equation for saturated hydrocarbons.

The regression equation may be rewritten to reveal a term that may be related to the degree of unsaturation in a molecule. Since

$$\Delta\chi_i^v = \chi_n^v - \chi_i^v \qquad \text{and} \qquad \Delta\chi_i = \chi_n - \chi_i$$

and because the coefficients of $\Delta\chi$ and $\Delta\chi^v$ are opposite in sign, the equation may be rewritten

$$\Delta H_a^i = a_0 + a_1(\chi_i - \chi_i^v) + a_2 \, \Delta\chi_i + a_3 n$$

The four-variable equation above is significantly better than the relation based solely on the number of carbon atoms n, for which $s = 2.34$ kcal. The addition of both $^1\chi$ and $^1\chi^v$ gives a relation almost as good as the four-variable equation, $s = 1.572$ kcal.

The best correlation achieved with these 37 alkenes using the $^m\chi_t^v$ terms plus $^1\chi$ gives a standard error of 1.089 kcal:

$$\Delta H_a = 38.99 + 22.64n - 24.14 \, {}^1\chi - 7.545 \, {}^2\chi^v - 5.673 \, {}^3\chi_P^v$$

$$- 10.46 \, {}^5\chi_C^v + 5.924 \, {}^4\chi_{PC}^v + 13.86 \, \chi_n^v$$

The predicted ΔH_a values are shown as the last column in Table XII.

Since the probable experimental error is about one-third the regression error, it appears that adjacency relations alone do not suffice for the treatment of ΔH_a values for olefins. Questions of thermodynamic stability are complex and may require additional descriptors.

It is tempting to analyze the regression terms with thermochemical data in terms of the structural characteristics described by each term in the connectivity function. At a superficial level of analysis we might say that $^1\chi$ is largely a count of bonds, while other path terms describe linearity and/or effects such as 1–4 and 1–5 *gauche* steric interactions. The $^m\chi_C$ cluster terms may reflect certain stabilizations or steric instabilities such as *gem* and *vic* dimethyl effects at branch points in the molecule. All of these suppositions are interesting but perhaps premature at this time. A large body of results together with careful dissection and analysis of each term is necessary for this exercise. It is anticipated that this will be accomplished by regular contributions of connectivity studies.

II. Molar Refraction and Molecular Polarizability

Refractive index measurements yield information about the ability of the molecular electron distribution to be deformed in an electric field or in the presence of other molecules. The molecular electronic polarizability α_E is then a measure of the polarizability of the electron cloud of the molecule.

Molar refraction R_m is defined as

$$R_m = \frac{n^2 - 1}{n^2 + 2} \frac{M}{d}$$

where M is the molecular weight, d the density, and n the refractive index. Although n and d depend on physical state, R_m does not. Thus, molar refraction is a molecular property whereas refractive index, density, and molar volume (M/d) are bulk or molar properties.

Refractive index is a unitless quantity. The units of R_m are volume units, usually cm^3 mol^{-1}. The molar refraction is a measure of the volume of the isolated molecule, whereas V_m $(=M/d)$ is the effective volume occupied by one mole in the liquid state.

Although the molecular volume aspect of R_m is of some interest to chemists and biologists, its relationship to polarizability is of special interest to drug design workers. Electronic polarization or molecular polarizability α_E and molar refraction are related as

$$R_m = 4/3\pi\alpha_E$$

The polarizability α of a molecule is related to the intermolecular forces important in the interaction of drug and receptor. Specifically, the α term is the dominant term in the expression for the dispersion energy of interaction. The use of α and R_m have become increasingly important recently in correlating physical properties with biological actions of drug molecule series [4].

The ability to predict R_m from a simple chemical graph could have significant consequences in gaining insight into the SAR of biological molecules as well as the study of physical properties such as heat of vaporization and boiling point.

TABLE XII

Heat of Atomization ΔH_a for Alkenes Correlated with Connectivity

Compound	$\Delta\chi^p$	χ_n^p	$\Delta\chi$	n	ΔH_a Obs	ΔH_a Calc
1. Ethylene	0.0000	0.5000	0.0000	2.0	537.75	540.43
2. Propylene	0.0000	0.9855	0.0000	3.0	820.42	818.45
3. 1-Butene	0.0000	1.5236	0.0000	4.0	1100.60	1102.61
4. *trans*-2-Butene	0.0355	1.5236	0.0000	4.0	1103.39	1102.19
5. 2-Methylpropene	0.1700	1.5236	0.1821	4.0	1104.66	1104.69
6. 1-Pentene	0.0000	2.0236	0.0000	5.0	1380.83	1382.32
7. *trans*-2-Pentene	−0.0024	2.0236	0.0000	5.0	1383.43	1382.34
8. 2-Methyl-1-butene	0.1093	2.0236	0.1441	5.0	1384.05	1384.26
9. 3-Methyl-1-butene	0.1273	2.0236	0.1441	5.0	1382.11	1384.05
10. 2-Methyl-2-butene	0.1575	2.0236	0.1441	5.0	1385.62	1383.70
11. 1-Hexene	0.0000	2.5236	0.0000	6.0	1660.55	1662.02
12. *trans*-2-Hexene	−0.0024	2.5236	0.0000	6.0	1663.48	1662.05
13. *trans*-3-Hexene	−0.0404	2.5236	0.0000	6.0	1663.61	1662.49
14. 2-Methyl-1-pentene	0.1093	2.5236	0.1441	6.0	1664.79	1663.97
15. 3-Methyl-1-pentene	0.0893	2.5236	0.1061	6.0	1662.43	1663.35

16.	4-Methyl-1-pentene	0.1441	2.5236	0.1441	6.0	1662.85	1663.56
17.	2-Methyl-2-pentene	0.1195	2.5236	0.1441	6.0	1666.58	1663.85
18.	3-Methyl-*trans*-2-pentene	0.0422	2.5236	0.1061	6.0	1665.69	1663.90
19.	4-Methyl-*trans*-2-pentene	0.1248	2.5236	0.1441	6.0	1665.30	1663.79
20.	2-Ethyl-1-butene	0.0487	2.5236	0.1061	6.0	1663.99	1663.82
21.	2,3-Dimethyl-1-butene	0.2266	2.5236	0.2714	6.0	1665.79	1665.44
22.	3,3-Dimethyl-1-butene	0.3266	2.5236	0.3535	6.0	1665.11	1666.11
23.	2,3-Dimethyl-2-butene	0.2736	2.5236	0.2714	6.0	1667.02	1664.90
24.	1-Heptene	0.0000	3.0236	0.0000	7.0	1940.51	1941.73
25.	5-Methyl-1-hexene	0.1093	3.0236	0.1441	7.0	1941.40	1943.68
26.	3-Methyl-*trans*-3-hexene	0.0589	3.0236	0.1061	7.0	1944.06	1943.42
27.	2,4-Dimethyl-1-pentene	0.2535	3.0236	0.2883	7.0	1945.73	1945.22
28.	4,4-Dimethyl-1-pentene	0.3535	3.0236	0.4059	7.0	1945.14	1946.68
29.	2,4-Dimethyl-2-pentene	0.2468	3.0236	0.2883	7.0	1946.90	1945.29
30.	4,4-Dimethyl-*trans*-2-pentene	0.3242	3.0236	0.4059	7.0	1946.92	1947.02
31.	3-Methyl-2-ethyl-1-butene	0.1660	3.0236	0.2334	7.0	1994.71	1945.01
32.	2,3-Trimethyl-1-butene	0.4200	3.0236	0.4059	7.0	1946.13	1945.90
33.	1-Octene	0.0000	3.5236	0.0000	8.0	2220.21	2221.44
34.	2,2-Dimethyl-*trans*-3-hexene	0.2862	3.5236	0.3535	8.0	2226.53	2226.00
35.	2-Methyl-3-ethyl-1-pentene	0.1506	3.5236	0.1954	8.0	2224.77	2224.05
36.	2,4,4-Trimethyl-2-pentene	0.4462	3.5236	0.4977	8.0	2225.87	2227.35
37.	1-Decene	0.0000	4.5236	0.0000	10.0	2780.48	2780.86

A.   First-Order Treatment of Mixed Classes of Compounds

In the first study reported on molecular connectivity [5] a series of molecules of diverse structures and atom types were analyzed. At this primitive level all first-row atoms were treated equivalently. The δ values were assigned on the basis of simple connectivity. The ring modification term, -0.50 per ring, was applied for all cyclic molecules.

The values of $^1\chi$ were found to correlate rather well with polarizability (see Table XIII):

$$\alpha = 1.60 - 9.26\ ^1\chi, \qquad r = 0.990, \quad s = 3.59, \quad N = 36$$

The correlation is significant. The standard error may be unsatisfactory for some applications but the quality of this correlation with a single variable is quite remarkable for the wide variety of compounds involved in this study.

With the use of valence δ values and the connectivity function, it is possible to examine the dependence of molar refraction on molecular connectivity at a greater level of sophistication. We shall look at several classes of compounds individually as well as in a single group.

B.   Alcohols*

For 31 aliphatic alcohols $^1\chi$ correlates well with R_m: $r = 0.9886$, $s = 1.02$. Correlation with $^1\chi^v$ is only slightly better: $r = 0.9926$ and $s = 0.92$. However, these two indices together provide a significantly better correlation:

$$R_m = 7.190 + 18.30\ ^1\chi^v - 8.914\ ^1\chi, \quad r = 0.9959, \quad s = 0.720, \quad N = 31$$

The use of additional extended terms leads to very significant improvement in the multiple regression. For the number of compounds studied, the maximum number of variables that we felt meaningful is four:

$$R_m = 3.518 + 9.409\ ^1\chi^v + 4.832\ ^3\chi_C^v - 0.436\ ^3\chi_P - 1.326\ ^3\chi_C$$

$$r = 0.9998, \qquad s = 0.153, \qquad N = 31$$

A relative error of about 0.3% is quite significant for molar refraction in this case. Propagation of error analysis** suggests that the experimental

---

* Unless otherwise stated, data in Chapters 4 and 5 are obtained from "Handbook of Tables for Organic Compound Identification" or "Handbook of Chemistry and Physics," 52nd ed., CRC Press, Cleveland, Ohio.

** For R_m values in the range 25.0–35.0 an examination of tabulated data suggests the following reasonable values for experimental error: refractive index n, $\sigma_n = 0.00005$ to 0.0005; density d, $\sigma_d = 0.00005$ to 0.0005; and molecular weight M, $\sigma_M = 0.00002$.

TABLE XIII

Correlation of Polarizability α with Connectivity Index for Local Anesthetics

Compound	$^1\chi$	α Obs	α Calc
Methanol	1.000	8.2	10.9
Ethanol	1.414	12.9	14.7
Acetone	1.732	16.2	17.7
Isopropanol	1.732	17.6	17.7
n-Propanol	1.914	17.5	19.3
Urethane	2.769	23.2	27.3
Ethyl ether	2.414	22.5	24.0
n-Butanol	2.414	22.1	24.0
Pyridine	2.500	24.1	24.8
Hydroquinone	3.288	29.4	32.1
Aniline	2.893	31.6	28.4
Benzyl alcohol	3.432	32.5	33.4
n-Pentanol	2.914	26.8	28.6
Phenol	2.893	27.8	28.4
Toluene	2.893	31.1	28.4
Benzimidazole	3.466	40.2	33.7
n-Hexanol	3.414	31.4	33.2
Nitrobenzene	3.804	32.5	36.8
Quinoline	3.966	42.1	38.3
8-Hydroxyquinoline	4.431	44.7	42.6
n-Heptanol	3.914	36.0	37.9
2-Naphthol	4.359	45.4	42.0
Methyl Anthranilate	4.752	48.9	45.6
n-Octanol	4.414	40.6	42.5
Thymol	4.608	47.3	44.3
o-Phenanthroline	5.448	57.8	52.1
Ephedrine	4.753	50.2	50.3
Procaine	7.668	67.0	72.6
Lidocaine	7.685	72.5	72.8
Diphenylhydramine	8.127	79.5	78.2
Tetracaine	8.629	79.7	81.5
Phenyltoloxamine	8.254	79.9	78.1
Quinine	9.705	93.8	91.5
Eserine	7.969	82.4	75.4
Caramiphen	9.226	87.0	87.0
Dibucaine	11.189	103.6	105.2

standard error lies in the range 0.007–0.05 cm^3 mol^{-1}. For a set of data taken from several sources, a regression error of 0.16 could be considered excellent. Table XIV shows R_m calculated by the multiple-regression equation above.

C. Ethers

Results comparable to those for alcohols are obtained for aliphatic ethers. Both $^1\chi$ and $^1\chi^v$ produce the same standard error: $r = 0.991$ and $s = 0.787$ for $^1\chi^v$. Using the two together produces no significant improvement:

$$R_m = 3.780 + 8.477 \,^1\chi, \qquad r = 0.987, \quad s = 0.784, \quad N = 9$$

A dramatic improvement occurs, however, with the addition of only one subgraph term, $^3\chi_C$:

$$R_m = 3.569 + 9.070 \,^1\chi^v + 1.953 \,^3\chi_C, \qquad r = 0.9989, \quad s = 0.291, \quad N = 9$$

Observed and calculated R_m values are shown in Table XV for the two-variable equation.

D. Amines

Correlation of molar refraction data for amines is not of the same high quality as for alcohols and ethers. Examination of the refractive index and density data reveals much greater variability in the quality of the data.

$^1\chi$ gives a standard error of 2.30 cm^3 mol^{-1}, whereas $^1\chi^v$ yields $r = 0.997$ and $s = 1.92$. The addition of extended terms produces some improvement:

$$R_m = 2.403 + 12.603 \,^1\chi^v + 2.602 \,^3\chi_P - 10.122 \,^3\chi_P^v$$

$$r = 0.9871, \qquad s = 1.333, \qquad N = 2^\frown$$

There appears to be a fundamental difference, with respect to molar refraction, between the three classes of amines. An additional variable is introduced to describe the electronic environment of the mono-, di-, and trisubstituted nitrogen atoms. The values of the atom connectivity δ_n^v are used since this number contains both valence electron count and the number of attached hydrogens:

$$R_m = 1.026 + 11.73 \,^1\chi^v - 5.778 \,^3\chi_P^v + 0.760\delta_N^v$$

$$r = 0.9881, \qquad s = 1.280, \qquad N = 22$$

TABLE XIV

Correlation of Alcohol Molar Refraction R_m with the Connectivity Function

Compound	R_m	
	Obs	Calc
1. 2-Propanol	17.705	17.293
2. 1-Propanol	17.529	17.633
3. 2-Methyl-1-propanol	22.103	22.274
4. 1-Butanol	22.067	22.247
5. 2-Methyl-2-butanol	26.722	26.795
6. 2-Pentanol	26.680	26.542
7. 3-Methyl-1-butanol	26.904	26.957
8. 2-Methyl-1-butanol	26.697	26.665
9. 1-Pentanol	26.801	26.843
10. 3-Pentanol	26.618	26.570
11. 2-Methyl-2-pentanol	31.211	31.526
12. 3-Methyl-3-pentanol	31.183	31.128
13. 4-Methyl-2-pentanol	31.351	31.288
14. 2-Methyl-3-pentanol	31.138	31.031
15. 4-Methyl-1-pentanol	31.489	31.544
16. 2-Methyl-1-pentanol	31.164	31.333
17. 2-Ethyl-1-butanol	31.180	31.284
18. 1-Hexanol	31.429	31.439
19. 2,4-Dimethyl-3-pentanol	35.675	35.531
20. 3-Ethyl-3-pentanol	35.822	35.691
21. 2-Methyl-1-hexanol	35.931	35.920
22. 1-Heptanol	36.094	36.034
23. 2-Methyl-2-heptanol	40.899	40.704
24. 3-Methyl-3-heptanol	40.447	40.428
25. 4-Methyl-4-heptanol	40.439	40.563
26. 6-Methyl-1-heptanol	40.737	40.736
27. 2-Ethyl-1-hexanol	40.625	40.524
28. 1-Octanol	40.638	40.630
29. 2,6-Dimethyl-4-heptanol	45.521	45.383
30. 2-Methyl-2-octanol	45.207	45.300
31. 4-Ethyl-4-heptanol	44.920	45.100

Table XVI contains observed R_m values and calculations based on just the preceding equation.

E. Halogen-Containing Saturated Compounds

The empirical valence δ^v values for halogens may be used in a study of the molar refraction of alkyl chlorides, bromides, and iodides. Because a

TABLE XV

Molar Refraction R_m of Aliphatic Ethers Correlated with the Connectivity Function

		R_m	
Compound		Obs	Calc
1.	Methylpropyl ether	22.0493	22.259
2.	Diethyl ether	22.4934	22.809
3.	*n*-Butylmethyl ether	27.0208	26.794
4.	*sec*-Butylethyl ether	31.5602	31.724
5.	*sec*-Butylmethyl-2-methyl ether	31.3370	31.419
6.	Di-*n*-propyl ether	32.2262	31.879
7.	Butylethyl ether	31.7335	31.879
8.	Butyl-isopropyl ether	36.0273	36.148
9.	Dibutyl ether	40.9872	40.949

TABLE XVI

Correlation of Molar Refraction R_m for Aliphatic Amines with the Connectivity Function

		R_m	
Compound		Obs	Calc
1.	Trimethyl amine	19.595	20.56
2.	1-Aminopropane	19.401	20.58
3.	2-Amino-2-methylpropane	24.257	24.28
4.	1-Aminobutane	24.079	24.89
5.	1-Amino-2,2-dimethylpropane	28.471	26.29
6.	1-Amino-3-methylbutane	28.672	27.99
7.	3-Aminopentane	28.617	27.32
8.	Dipropyl amine	33.515	35.74
9.	1-Aminopentane	28.728	29.31
10.	3-Amino-2,2-dimethylbutane	25.098	27.07
11.	Di-isopropyl amine	33.641	34.07
12.	Butyldimethyl amine	33.816	33.50
13.	Triethyl amine	33.794	33.08
14.	Butylethyl amine	33.452	33.09
15.	1-Aminohexane	33.290	33.73
16.	Dimethylpentyl amine	38.281	37.92
17.	2-Aminoheptane	38.038	37.36
18.	1-Aminoheptane	38.003	38.15
19.	Di-isobutyl amine	42.920	42.63
20.	Dimethyl-isobutyl amine	33.852	33.05
21.	Tripropyl amine	47.783	49.07
22.	1-Aminononane	47.277	46.99

104

mixed set of compounds is used, $^1\chi$ does not provide a good correlation, giving $r = 0.917$ and $s = 3.10$. $^1\chi^v$ is significantly better with $r = 0.985$ and $s = 1.33$. Using the two together is only slightly better:

$$R_m = 5.766 + 2.515 \, ^1\chi + 5.678 \, ^1\chi^v, \quad r = 0.9888, \quad s = 1.198, \quad N = 17$$

The use of a five-term connectivity function produces a very excellent correlation with a standard error approaching the experimental error:

$$R_m = 2.577 + 4.256 \, ^1\chi + 4.832 \, ^1\chi^v + 1.598 \, ^3\chi_C + 0.314 \, ^3\chi_C^v + 1.668 \, ^3\chi_P$$

$$r = 0.999, \quad s = 0.1080, \quad N = 17$$

Observed and calculated results are given in Table XVII.

Such excellent results are another corroboration of the usefulness of the δ^v values. No dummy variables are introduced to distinguish between the various halogen atoms.

F. A Study of Mixed Classes of Compounds and the Connectivity Function

To test further the usefulness and validity of the δ^v values, a mixed set of alcohols, ethers, primary and secondary amines, and three types of

TABLE XVII

Correlation of Molar Refraction R_m for Mixed Alkyl Halides with the Connectivity Function

Compound	R_m	
	Obs	Calc
1. 1-Chloropropane	20.847	20.85
2. 2-Chlorobutane	25.506	25.53
3. 1-Chloro-2-methylpropane	25.360	25.54
4. 2-Chlorobutane	25.441	25.45
5. 3-Chloropentane	30.161	30.19
6. 2-Bromopropane	23.935	23.95
7. 1-Bromopropane	23.679	23.70
8. 2-Bromobutane	28.651	28.46
9. 1-Bromo-2-methylpropane	28.537	28.39
10. 1-Bromobutane	28.347	28.26
11. 3-Bromopentane	33.068	33.00
12. 2-Iodobutane	33.940	33.92
13. 3-Iodopentane	38.354	38.38
14. 2-Iodopentane	38.314	38.42
15. 1-Iodopentane	38.264	38.30
16. 1-Iodohexane	42.891	42.92
17. 1-Iodoheptane	47.610	47.55

TABLE XVIII

Correlation of Molar Refraction R_m of Mixed Aliphatic Compounds with the Connectivity Function

Compound	R_m	
	Obs	Calc
1. 2-Propanol	17.705	17.642
2. 2-Methyl-1-propanol	22.103	22.296
3. 2-Methyl-2-butanol	26.722	26.955
4. 2-Pentanol	26.681	26.754
5. 3-Methyl-1-butanol	26.904	26.791
6. 2-Methyl-1-butanol	26.697	26.855
7. 3-Pentanol	26.618	26.941
8. 2-Methyl-2-pentanol	31.211	31.450
9. 2,2-Dimethyl-1-butanol	31.269	31.800
10. 3-Methyl-3-pentanol	31.183	31.431
11. 4-Methyl-2-pentanol	31.351	30.899
12. 2-Methyl-3-pentanol	31.138	30.994
13. 4-Methyl-1-pentanol	31.489	31.286
14. 2-Methyl-1-pentanol	31.164	31.350
15. 2-Ethyl-1-butanol	31.180	31.496
16. 1-Hexanol	31.429	31.636
17. 3-Ethyl-3-pentanol	35.822	36.032
18. 2-Methyl-1-hexanol	35.931	35.845
19. 1-Heptanol	36.094	36.131
20. 3-Methyl-3-heptanol	40.447	40.421
21. 4-Methyl-4-heptanol	40.440	40.421
22. 6-Methyl-1-heptanol	40.737	40.276
23. 2-Ethyl-1-hexanol	40.625	40.486
24. 1-Octanol	40.638	40.626
25. 4-Ethyl-4-heptanol	44.920	45.022
26. Ethyl-isopropyl ether	27.679	26.077
27. Butylmethyl ether	27.021	27.338
28. *sec*-Butylmethyl-2-methyl ether	31.337	31.763
29. *sec*-Butylethyl ether	31.560	31.854
30. Dipropyl ether	32.226	32.153
31. Butyl-isopropyl ether	36.027	36.220
32. Ethylpentyl ether	36.364	36.393
33. Dibutyl ether	40.987	41.143
34. Trimethyl amine	19.595	19.062
35. 1-Aminopropane	19.401	19.999
36. 1-Aminobutane	24.079	24.494
37. 1-Amino-3-methylbutane	28.672	28.638
38. 3-Aminopentane	28.617	28.768
39. 1-Aminopentane	28.728	28.989
40. Butyldimethyl amine	33.816	33.590

TABLE XVIII—Continued

| | R_m | |
Compound	Obs	Calc
41. Triethyl amine	33.794	34.288
42. 1-Aminohexane	33.290	33.484
43. Dimethylpentyl amine	38.281	38.085
44. 2-Aminoheptane	38.038	37.601
45. 1-Aminoheptane	38.003	37.979
46. Tripropyl amine	47.784	47.773
47. 1-Aminononane	47.277	46.969
48. 1-Chloropropane	20.847	20.974
49. 2-Chlorobutane	25.506	25.132
50. 1-Chloro-2-methylpropane	25.360	25.119
51. 2-Chlorobutane	25.441	25.469
52. 3-Chloropentane	30.161	29.647
53. 2-Bromopropane	23.935	24.021
54. 1-Bromopropane	23.679	23.944
55. 1-Bromo-2-methylpropane	28.537	28.089
56. 2-Bromobutane	28.651	28.155
57. 1-Bromobutane	28.347	28.439
58. 3-Bromopentane	33.068	32.495
59. 2-Iodobutane	33.940	33.985
60. 2-Iodopentane	38.314	38.480
61. 3-Iodopentane	38.354	37.987
62. 1-Iodopentane	38.264	38.663
63. 1-Iodohexane	42.891	43.158
64. 1-Iodoheptane	47.610	47.653

halides was correlated with the connectivity function [6]. As expected for such a mixed set of compounds, correlation with $^1\chi$ yields a high standard error, 3.33, and correlation with $^1\chi^v$ gives a standard error of 2.30. The two together, however, give a better fit at about the 2% level of relative error:

$$R_m = 4.455 + 3.695\ ^1\chi + 5.208\ ^1\chi^v, \quad r = 0.9897, \quad s = 1.01, \quad N = 64$$

The use of a five-variable function reduces the standard error to 0.364:

$$R_m = 3.004 + 3.714\ ^1\chi + 5.276\ ^1\chi^v + 0.478\ ^3\chi_C + 1.839\ ^3\chi_C^v + 0.454\delta_N^v$$

$$r = 0.9987, \quad s = 0.364, \quad N = 64$$

Values for R_m predicted for this equation are shown in Table XVIII. The electronic character of the nitrogen atoms in the substituted amines seems to require the δ_N^v variable, whereas the δ_O^v variable, added to the equation for the alcohols and ethers, does not significantly improve the correlation. This high-quality correlation is more encouraging for a set of data including heteroatomic molecules with nitrogen, oxygen, chlorine, bromine, and iodine.

G. Hydrocarbons

Three classes of hydrocarbons have been selected for study because of the availability of very good data for a significant number of compounds. These classes are alkanes, olefins with one double bond, and variously substituted alkyl benzenes.

1. Alkanes

The generally excellent data for alkanes are frequently used as a critical test for a new method. There are available for 46 alkanes density data with five significant figures and refractive index data with six significant figures. On the basis of error propagation analysis we estimate the lower limit on the experimental error on R_m to be 0.007 cm³ mol⁻¹.

Correlation with $^0\chi$ is excellent: $r = 0.9848$, whereas $^1\chi$ provides a lesser correlation:

$$R_m = 1.369 + 10.440\,^1\chi, \qquad r = 0.952, \quad s = 1.81, \quad N = 46$$

Although correlation with the number of carbon atoms n is also excellent, n is not required along with the connectivity function for a fit of exceptional quality. The use of $^1\chi$ and $^2\chi$ together drops the standard error to 0.121:

$$R_m = 3.707 + 7.756\,^1\chi + 2.207\,^2\chi, \qquad r = 0.9997, \quad s = 0.121, \quad N = 46$$

TABLE XIX

Correlation of Alkane Molar Refraction with the Connectivity Function

Statistical data		Regression coefficients					
r	s	$^1\chi$	$^2\chi$	$^3\chi_P$	$^4\chi_C$	$^4\chi_{PC}$	Constant
0.9522	1.806	10.440					1.369
0.9997	0.122	7.756	2.207				3.707
0.9999	0.0738	7.615	2.198	0.2035			3.883
0.9999	0.0451	7.528	2.325	0.1979	−0.5983		3.884
0.9999	0.0273	7.7331	2.423	0.4540	−0.6185	−0.1409	4.008

TABLE XX *Correlation of Molar Refraction R_m of Alkanes with the Connectivity Function*

Compound	R_m	
	Obs	Calc
1. n-Pentane	25.267	25.307
2. 2-Methylbutane	25.294	25.329
3. n-Hexane	29.981	29.943
4. 3-Methylpentane	29.949	29.956
5. 2-Methylpentane	29.804	29.801
6. 2,2-Dimethylbutane	29.938	29.954
7. 2,3-Dimethylbutane	29.813	29.828
8. n-Heptane	34.555	34.579
9. 2-Methylhexane	34.595	34.600
10. 3-Methylhexane	34.464	34.439
11. 3-Ethylpentane	34.287	34.304
12. 2,2-Dimethylpentane	34.621	34.597
13. 2,3-Dimethylpentane	34.328	34.327
14. 2,4-Dimethylpentane	34.623	34.611
15. 3,3-Dimethylpentane	34.336	34.321
16. 2,2,3-Trimethylbutane	34.378	34.358
17. n-Octane	39.194	39.214
18. 2-Methylheptane	39.234	39.236
19. 3-Methylheptane	39.102	39.083
20. 4-Methylheptane	39.119	39.077
21. 3-Ethylhexane	38.946	38.954
22. 2,2-Dimethylhexane	39.255	39.246
23. 2,3-Dimethylhexane	38.983	38.969
24. 2,4-Dimethylhexane	39.132	39.099
25. 2,5-Dimethylhexane	39.261	39.261
26. 3,3-Dimethylhexane	39.011	38.969
27. 3,4-Dimethylhexane	38.864	38.830
28. 2-Methyl-3-ethylpentane	38.838	38.848
29. 3-Methyl-3-ethylpentane	38.719	38.725
30. 2,2,3-Trimethylpentane	38.927	38.882
31. 2,2,4-Trimethylpentane	39.264	39.260
32. 2,3,3-Trimethylpentane	38.764	38.752
33. 2,3,4-Trimethylpentane	38.870	38.866
34. n-Nonane	43.846	43.850
35. 2,2,5-Trimethylhexane	43.939	43.910
36. 2,4,4-Trimethylhexane	43.663	43.634
37. 3,3-Diethylpentane	43.117	43.157
38. 2,2,3,3-Tetramethylpentane	43.218	43.234
39. 2,2,3,4-Tetramethylpentane	43.439	43.435
40. 2,2,4,4-Tetramethylpentane	43.878	43.916
41. 2,3,3,4-Tetramethylpentane	43.205	43.198
42. 2,4-Dimethyl-3-i-propylpentane	47.913	47.968
43. 2,2,4,5-Tetramethylhexane	48.262	48.284
44. 2,2,5,5-Tetramethylhexane	48.575	48.559
45. 2,2,3,4,4-Pentamethylpentane	47.988	48.014
46. 2,2,3,3-Tetramethylhexane	47.905	47.886

Additional subgraph terms continue to improve the quality. We report here a five-variable equation. Additional statistical data are in Table XIX:

$$R_m = 4.008 + 7.331\ ^1\chi + 2.423\ ^2\chi + 0.454\ ^3\chi_P - 0.619\ ^4\chi_C - 0.141\ ^4\chi_{PC}$$

$$r = 0.9999, \qquad s = 0.027, \qquad N = 46$$

Observed and calculated values are given in Table XX. The experimental error is estimated to lie between 0.007 and 0.04.

The standard error reported here is very low, approximating the estimated standard experimental error. It appears then that molecular connectivity provides an excellent model for molar refraction and polarization in alkanes. The results reported earlier in this section support the practical application of this method for heteroatomic molecules as well. Since intermolecular forces in organic molecules frequently involve polarizability, the molecular connectivity method may be useful in those areas which relate to such forces.

2. *Monoolefinic Alkenes*

For the set of alkenes for which there is good data, $^1\chi$ yields a correlation coefficient of $r = 0.954$ and $s = 1.24$. The addition of $^2\chi$ and $^0\chi^v$ decreases the error by more than a factor of five:

$$R_m = 6.686 + 0.526\ ^0\chi^v + 6.905\ ^1\chi^v + 1.935\ ^2\chi$$

$$r = 0.9985, \qquad s = 0.223, \qquad N = 39$$

This three-variable function reproduces the R_m values with 0.7% relative error. The addition of other subgraph terms such as $^5\chi_C$ and $^1\chi^v$ improves the relationship somewhat, giving $s = 0.204$. Observed and predicted values are shown in Table XXI, based on the six-variable equation

$$R_m = 4.089 - 0.0017\ ^0\chi^v + 6.611\ ^1\chi + 0.650\ ^1\chi^v$$

$$+ 2.444\ ^2\chi + 0.479\ ^3\chi_P^v - 0.924\ ^4\chi_C$$

$$r = 0.9994, \qquad s = 0.147, \qquad N = 39$$

The relative error here is 0.4%.

3. *Alkyl-Substituted Benzenes*

This set of data includes 70 compounds ranging from benzene to compounds with substituents containing six carbon atoms with a wide variety of branching and substitution patterns. The best correlation with a single variable is with $^1\chi^v$: $r = 0.984$, $s = 0.97$. The addition of other

TABLE XXI

Observed and Calculated Molar Refraction R_m of Alkenes Correlated with the Connectivity Function

		R_m	
Compound		Obs	Calc
1.	2-Methyl-2-butene	24.955	24.983
2.	1-Pentene	24.858	24.904
3.	3,3-Dimethyl-1-butene	29.598	29.529
4.	2,3-Dimethyl-2-butene	29.590	29.575
5.	2,3-Dimethyl-1-butene	30.063	29.600
6.	4-Methyl-1-pentene	29.542	29.582
7.	2-Methyl-1-pentene	29.398	29.623
8.	2-Methyl-2-pentene	29.754	29.586
9.	3-Methyl-1-pentene	29.485	29.426
10.	2-Ethyl-1-butene	29.391	29.431
11.	1-Hexene	29.208	29.526
12.	2,2,3-Trimethyl-1-butene	33.980	34.184
13.	4,4-Dimethyl-1-pentene	34.233	34.178
14.	3,3-Dimethyl-1-pentene	34.011	33.997
15.	2,3-Dimethyl-2-pentene	34.203	34.025
16.	3,4-Dimethyl-1-pentene	33.919	34.042
17.	3,4-Dimethyl-2-pentene	33.900	34.016
18.	2-Ethyl-3-methyl-1-butene	34.024	34.002
19.	2,3-Dimethyl-1-pentene	34.005	34.075
20.	5-Methyl-1-hexene	34.139	34.223
21.	2-Methyl-2-hexene	34.395	34.188
22.	2-Methyl-1-hexene	34.114	34.250
23.	2-Methyl-3-hexene	34.223	34.169
24.	4-Methyl-1-hexene	34.078	34.079
25.	2-Ethyl-1-pentene	33.991	34.039
26.	3-Ethyl-1-pentene	34.039	33.905
27.	3-Ethyl-2-pentene	34.117	33.883
28.	1-Heptene	34.136	34.138
29.	2,4,4-Trimethyl-2-pentene	39.015	38.851
30.	2,4,4-Trimethyl-1-pentene	38.769	38.893
31.	2,3,4-Trimethyl-1-pentene	38.836	38.655
32.	3,3,4-Trimethyl-1-pentene	38.499	38.609
33.	2-Isopropyl-3-methyl-1-butene	38.385	38.569
34.	5,5-Dimethyl-1-hexene	38.785	38.853
35.	4,4-Dimethyl-1-hexene	38.643	38.668
36.	3,3-Dimethyl-1-hexene	38.548	38.568
37.	2,5-Dimethyl-3-hexene	38.823	38.848
38.	3,5-Dimethyl-1-hexene	38.764	38.719
39.	3-Ethyl-4-methyl-1-pentene	38.591	38.513

TABLE XXII

Statistical Data for Correlation of Molar Refraction R_m of Alkyl Benzenes with the Connectivity Function

Statistical data		Regression coefficients				
r	s	$^1\chi^v$	$^2\chi^v$	$^1\chi$	$^2\chi$	Constant
0.9841	0.966	9.492				8.944
0.9987	0.270	7.548			2.501	6.265
0.9995	0.166	7.754	−2.662		4.886	3.461
0.9995	0.163	4.844	−2.932	2.972	5.157	0.094

terms decreases the standard error significantly. The following four-variable function produces the best correlation:

$$R_m = 2.971\ ^1\chi + 4.844\ ^1\chi^v + 5.157\ ^2\chi - 2.932\ ^2\chi^v + 0.942$$

$$r = 0.9995, \qquad s = 0.163, \qquad N = 70$$

This standard error corresponds to 0.35% relative error. Statistical data are summarized in Table XXII; observed and calculated values are given in Table XXIII.

Two valence terms are used in valence/nonvalence pairs: $^1\chi/^1\chi^v$ and $^2\chi/^2\chi^v$. The second pair also occurs with coefficients of opposite sign.

A correlation study was run on the combined alkenes and alkyl benzenes, a set of 117 compounds. The best single variable is found to be $^0\chi$: $r = 0.981$, $s = 1.58$. Addition of $^1\chi$ decreases s to 1.034. The standard error decreases with the addition of subgraph terms. The statistical results are tabulated in Table XXII. A seven-variable equation reduces the standard error to 0.429:

$$R_m = 0.690\ ^0\chi + 0.717\ ^0\chi^v + 5.70\ ^1\chi - 3.43\ ^5\chi_P + 1.39\ ^6\chi_P$$

$$+ 1.68\ ^2\chi^v + 4.99\ ^3\chi_C^v + 4.81$$

$$r = 0.9986, \qquad s = 0.429, \qquad N = 117$$

III. Gas Equation Empirical Constants

Empirical equations that describe the pressure–volume–temperature behavior of real gases contain parameters characteristic of the gas molecules. These quantities relate to the volume occupied by the molecules and to the forces of interaction between molecules in the gas phase. These factors are often described as causing deviations from ideal behavior as described by the familiar equation

$$P = RT/V$$

where P is the pressure in atmospheres, T the temperature in degrees Kelvin, and V the molar volume in liters. R is the molar gas constant, 0.08205 liters-atm/deg mol.

In the van der Waals equation

$$P = RT/(V - b) - a/V^2$$

the quantity b takes account of the finite volume of the molecules, neglected in the ideal gas equation:

$$b = 4N_A v_m$$

where v_m is the molecular volume and N_A Avogadro's number. The quantity a, sometimes related to the molar attraction constant, is a measure of the attraction between molecules in the gas phase.

Ostrenga [7] and Mullins [8] have shown how the molar attraction constant can be used as a regression variable in drug studies. Other authors have shown that the molar attraction constant can be used in addition to or as a substitute for the water/octanol partition coefficient [9]. Cammarata *et al.* [10] have shown that these two quantities are related to each other and to molecular polarizability. The square of the molar attraction constant, the van der Waals constant a, and molecular polarizability are all proportional:

$$F^2 \sim a \sim \alpha \sim \log P_{oct/water}$$

The conditions under which these relations hold are restricted to substances that form regular solutions. Deviations from this relation may reveal something of the solution and partitioning process in biological systems.

A. Van der Waals Constants

It appears then that certain gas equation constants may be described in molecular connectivity terms. The van der Waals constant a for a set of mixed compounds requires only the simple connectivity index $^1\chi$ for adequate correlation, as shown in Table XXIV. The regression equation is given as

$$a = 2.88 + 6.38 \, ^1\chi, \qquad r = 0.987, \quad s = 0.93, \quad N = 20$$

TABLE XXIII

Observed and Calculated Molar Refraction R_m for Alkylbenzenes Correlated with the Connectivity Function

		R_m	
Compound		Obs	Calc
1.	Benzene	26.188	26.212
2.	Toluene	31.058	31.045
3.	*m*-Xylene	35.968	35.928
4.	*p*-Xylene	36.011	36.118
5.	*o*-Xylene	35.807	35.320
6.	Ethylbenzene	35.776	35.814
7.	1,3,5-Trimethylbenzene	40.822	40.857
8.	1,2,4-Trimethylbenzene	40.699	40.439
9.	1,2,3-Trimethylbenzene	40.459	40.107
10.	Isopropylbenzene	40.430	40.301
11.	1-Ethyl-3-methylbenzene	40.660	40.660
12.	1-Ethyl-4-methylbenzene	40.707	40.649
13.	1-Ethyl-2-methylbenzene	40.455	40.399
14.	Propylbenzene	40.458	40.559
15.	*tert*-Butylbenzene	45.000	44.824
16.	1,2,3,5-Tetramethylbenzene	45.302	45.036
17.	1,2,3,4-Tetramethylbenzene	45.086	44.656
18.	1-Isopropyl-4-methylbenzene	45.533	45.135
19.	1-Isopropyl-3-methylbenzene	45.311	45.183
20.	1-Isopropyl-2-methylbenzene	45.086	44.918
21.	1,3-Dimethyl-5-ethylbenzene	45.508	45.626
22.	1,3-Dimethyl-4-ethylbenzene	45.345	45.280
23.	1,4-Dimethyl-2-ethylbenzene	45.337	45.280
24.	1,2-Dimethyl-4-ethylbenzene	45.385	45.207
25.	1,3-Dimethyl-2-ethylbenzene	45.143	45.021
26.	1,2-Dimethyl-3-ethylbenzene	45.131	44.949
27.	Isobutylbenzene	45.208	45.038
28.	1-Methyl-3-propylbenzene	45.358	45.441
29.	1-Methyl-4-propylbenzene	45.362	45.393
30.	*sec*-Butylbenzene	45.036	44.907
31.	1-Methyl-2-propylbenzene	45.138	45.143
32.	1,3-Diethylbenzene	45.353	45.464
33.	1,4-Diethylbenzene	45.402	45.417
34.	1,2-Diethylbenzene	45.131	45.240
35.	Butylbenzene	45.106	45.299
36.	1-Methyl-3-*tert*-butylbenzene	49.895	49.706
37.	1-Methyl-4-*tert*-butylbenzene	49.932	49.658
38.	1,2-Dimethyl-4-isopropylbenzene	50.071	49.695
39.	1,4-Dimethyl-2-isopropylbenzene	49.992	49.800
40.	1,3-Dimethyl-2-isopropylbenzene	49.743	49.573

TABLE XXIII—Continued

		R_m	
Compound		Obs	Calc
41.	1,2-Dimethyl-3-isopropylbenzene	49.773	49.467
42.	*tert*-Pentylbenzene	49.788	49.303
43.	1-Isobutyl-3-methylbenzene	50.115	49.920
44.	1-Isobutyl-4-methylbenzene	50.104	49.872
45.	1-Isobutyl-2-methylbenzene	49.864	49.622
46.	(1'Methyl)-isobutylbenzene	48.924	49.291
47.	1-Ethyl-3-isopropylbenzene	50.379	49.951
48.	1-Methyl-3-*sec*-butylbenzene	49.962	49.788
49.	1-Ethyl-4-isopropylbenzene	50.132	49.904
50.	1-Methyl-4-*sec*-butylbenzene	49.833	49.741
51.	1,4-Dimethyl-2-ethylbenzene	50.019	50.015
52.	1,3-Dimethyl-4-ethylbenzene	49.956	50.025
53.	1-Ethyl-2-isopropylbenzene	49.773	49.759
54.	1-Methyl-2-*sec*-butylbenzene	49.699	49.524
55.	1,3-Diethyl-5-methylbenzene	50.266	50.394
56.	1,2-Diethyl-4-methylbenzene	50.099	50.122
57.	2,4-Diethyl-1-methylbenzene	50.078	50.049
58.	Isopentylbenzene	49.628	49.776
59.	1,2-Diethyl-3-methylbenzene	49.811	49.863
60.	1-Butyl-3-methylbenzene	49.990	50.181
61.	1-Butyl-4-methylbenzene	50.020	50.133
62.	*sec*-Pentylbenzene	49.777	49.698
63.	1-Ethyl-3-propylbenzene	50.064	50.209
64.	(1'Ethyl)propylbenzene	49.365	49.606
65.	1-Ethyl-2-propylbenzene	49.805	49.985
66.	Pentylbenzene	49.742	50.039
67.	1,3-Di-isopropylbenzene	54.660	54.438
68.	1,2-Di-isopropylbenzene	54.055	54.279
69.	Hexylbenzene	54.392	54.438

It is typical of properties related to polarizability that $^1\chi^v$ does not improve the correlation when only two variables are used. Additional subgraph terms would likely improve the relationship when tested against a longer list of compounds.

The van der Waals volume term b does require the valence delta for satisfactory correlation as shown in Table XXIV. The regression equation is

$$b = 0.0327 + 0.0410\ ^1\chi - 0.0382\ ^1\chi^v, \quad r = 0.981, \quad s = 0.0071, \quad N = 20$$

TABLE XXIV

Observed and Calculated Values of van der Waals Constants a and b for a Selected Set of Compounds Correlated with the Connectivity Function

Compound	$^1\chi$	$^1\chi^v$	a Obs	a Calc	b Obs	b Calc
1. Acetone	1.732	1.204	13.91	13.93	0.0994	0.0577
2. Isobutanol	2.270	1.879	17.03	17.37	0.1143	0.1186
3. Chloroform	1.732	2.085	15.17	13.94	0.1022	0.0957
4. Diethylamine	2.414	2.121	19.15	18.29	0.1392	0.1236
5. Ethylacetate	2.770	1.904	20.45	20.57	0.1412	0.1390
6. Ethyl alcohol	1.414	1.023	12.02	11.91	0.0841	0.0868
7. Ethyl amine	1.414	1.115	10.60	11.91	0.0841	0.0864
8. Ethylchloride	1.414	1.558	10.91	11.91	0.0865	0.0848
9. Diethylether	2.414	1.992	17.38	18.30	0.1344	0.1241
10. Ethylbutyrate	4.308	2.965	30.07	30.38	0.1919	0.1979
11. Diethylsulfide	2.414	2.870	18.75	18.29	0.1214	0.1207
12. 1,2-Dichloroethylene	1.914	1.732	16.91	15.10	0.1086	0.1046
13. Methyl alcohol	1.000	0.447	9.52	9.27	0.0670	0.0720
14. n-Propylalcohol	1.914	1.523	14.92	15.10	0.1019	0.1054
15. 1,2-Dibromoethylene	1.914	2.624	13.98	15.10	0.0866	0.1012
16. Isobutylacetate	4.126	3.198	28.50	29.22	0.1833	0.1896
17. Ethylproprionate	3.308	2.465	24.39	24.00	0.1615	0.1589
18. n-Propylacetate	3.270	2.404	24.63	23.76	0.1619	0.1576
19. n-Propylamine	1.914	1.523	14.99	15.10	0.1090	0.1054
20. Propionitrile	1.914	1.284	16.44	15.10	0.1064	0.1063

B. Constants from the Beattie–Bridgeman Equation

The Beattie–Bridgeman equation is a more sophisticated gas law that describes *PVT* behavior to within 0.5% over a wide range of conditions [11]. The equation is written

$$P = [RT(1 - \epsilon)/V^2][V + B] - A/V^2$$

where *P*, *V*, *T*, and *R* have their usual meanings and the equation constants ϵ, *A*, and *b* are defined in terms of five empirical constants:

$$A = A_0(1 - a/V), \qquad B = B_0(1 - b/V), \qquad \epsilon = c/VT^3$$

The terms B_0 and *b* loosely relate to molecular volume, whereas A_0 and *a* are associated with intermolecular forces. These four quantities have been correlated with $^1\chi$ for a set of eleven hydrocarbons including three olefins. See Table XXV for observed values and chi terms. The regression equations and statistical data are as follows:

$$A_0 = -4.73 + 14.62\chi, \qquad r = 0.94, \quad s = 5.14, \quad N = 11$$

$$a = 0.0118 + 0.0553\chi, \qquad r = 0.95, \quad s = 0.015, \quad N = 11$$

$$B_0 = -0.0135 + 0.0614\chi, \qquad r = 0.94, \quad s = 0.021, \quad N = 11$$

$$b = 0.0273 + 0.1826\chi, \qquad r = 0.94, \quad s = 0.066, \quad N = 11$$

Since data are so limited for these constants, multiple regression seems unwarranted.

IV. Diamagnetic Susceptibility

It is known that all forms of matter possess magnetic properties to some degree. Paramagnetic substances are attracted into a magnetic field, whereas diamagnetic substances like typical organic materials are repelled by magnetic fields.

In the general method of determining diamagnetic susceptibility, the Gouy balance is used. The sample weight is determined in the presence and absence of a homogeneous magnetic field. The apparent mass difference is proportional to the susceptibility. The sample magnetization I is proportional to field strength H:

$$I = kH$$

TABLE XXV

Observed Values for Constants from the Beattie–Bridgeman Equation

	Compound	A_0	a	B_0	b	$c \times 10^{-4}$	${}^1\chi^v$
1.	Methane	2.277	0.01855	0.05587	−0.01587	12.83	0.000
2.	Ethane	5.880	0.05861	0.09400	0.01915	90.00	1.000
3.	Propane	11.920	0.07321	0.18100	0.04293	120.00	1.414
4.	n-Butane	17.794	0.12161	0.24620	0.09423	350.00	1.914
5.	Isobutane	16.604	0.11171	0.23540	0.07697	300.00	1.732
6.	n-Pentane	28.260	0.15099	0.39400	0.13960	400.00	2.414
7.	Neopentane	23.330	0.15174	0.33560	0.13358	400.00	2.000
8.	n-Heptane	54.520	0.20066	0.70816	0.19179	400.00	3.414
9.	Ethylene	6.152	0.04964	0.12156	0.03597	22.68	0.500
10.	1-Butene	16.698	0.11988	0.24046	0.10690	300.00	1.526
11.	Iso-1-butene	16.960	0.10860	0.24200	0.08750	250.00	1.354

TABLE XXVI

Statistical Data for the Connectivity Subgraph Function Correlation with Alkane Magnetic Susceptibility χ_M

Statistical data								Regression Coefficients						
r	s	n	$^0\chi$	$^1\chi$	$^2\chi$	$^3\chi_P$	$^4\chi_P$	$^5\chi_C$	$^4\chi_{PC}$	$^5\chi_{PC}$	Constant
0.9991	0.869	11.389	—								6.930
0.9783	4.277	—	—	21.650							16.713
0.9995	0.640	13.240	—	-3.652							5.766
0.9981	1.300	—	—	18.578	6.301						9.988
0.9998	0.415	—	11.670	6.679	-0.784						0.105
0.9999	0.392	—	—	1.524	6.089	3.181	4.259	-3.739	2.427	-2.003	14.451

TABLE XXVII

Observed and Predicted Values of Molar Magnetic Susceptibilities χ_M for Alkanes

Compound	χ_M	
	Obs	Calc
1. Propane	40.50	40.37
2. Isobutane	51.70	51.46
3. *n*-Pentane	63.05	63.30
4. 2-Methylbutane	64.40	63.67
5. 2,2-Dimethylpropane	63.10	63.26
6. *n*-Hexane	74.60	75.50
7. 2-Methylpentane	75.26	75.36
8. 3-Methylpentane	75.52	75.50
9. 2,2-Dimethylbutane	76.24	75.52
10. 2,3-Dimethylbutane	76.22	76.17
11. *n*-Heptane	85.24	85.82
12. 2-Methylhexane	86.24	86.30
13. 2,2-Dimethylpentane	86.97	87.51
14. 2,3-Dimethylpentane	87.51	87.48
15. 2,4-Dimethylpentane	87.48	86.83
16. 2,2,3-Trimethylbutane	88.36	88.23
17. *n*-Octane	96.63	97.14
18. 3-Methylheptane	97.99	98.00
19. 2,3-Dimethylhexane	98.77	99.00
20. 2,5-Dimethylhexane	98.15	98.04
21. 3,4-Dimethylhexane	99.06	98.87
22. 2,2,3-Trimethylpentane	98.86	99.05
23. 2,2,4-Trimethylpentane	98.34	98.26
24. *n*-Nonane	108.13	108.46
25. 4-Methyloctane	109.63	109.56
26. *n*-Decane	119.74	119.78
27. *n*-Undecane	131.84	131.10

The quantity k is referred to as the magnetic susceptibility and is dependent on physical state. The molar magnetic susceptibility χ_M,* however, is independent of physical state:

$$\chi_M = Mk/d$$

where M is the molecular weight and d the density. χ_M is thus a molecular property.

* This symbol is not to be confused with the connectivity terms ${}^m\chi_t$.

A. Hydrocarbons

Applicability of the connectivity concept has been tested on a set of 27 saturated hydrocarbons. The number of carbon atoms provides good correlation with susceptibility, with $r = 0.9991$, whereas $^1\chi$ alone gives unsatisfactory correlation. However, the addition of $^1\chi$ to the regression equation with n provides for a 25% decrease in the standard error from 0.87 to 0.64 as shown in Table XXVI.

Use of the connectivity function provides the basis for an excellent correlation. Use of the first three terms alone, $^0\chi$, $^1\chi$, $^2\chi$, gives $r = 0.9999$ and a standard error of 0.415, less than 0.5% relative error. This is an excellent result for only three regression variables. The addition of more connectivity function terms decreases the error but because of the limited number of data points, meaningful examination of other terms is not possible at this time. Table XXVII contains observed values of molar magnetic susceptibility along with predictions based on the last line of Table XXVI.

B. Aliphatic Alcohols

Reliable data for heteroatomic molecules are somewhat limited. A set of 15 alcohols will be used as a representative application to the heteroatomic case.

Correlation data are summarized in Table XXVIII. The use of $^1\chi$ or $^1\chi^v$ alone is unsatisfactory as is their combination. However, the addition of the valence cluster term $^3\chi_C^v$ improves the relationship dramatically. Use of $^1\chi^v$ and $^3\chi_C^v$ with or without $^0\chi$ gives a standard

TABLE XXVIII

Statistical Data for Correlation of Aliphatic Alcohol Molar Magnetic Susceptibility χ_M with the Connectivity Function

Statistical data		Regression coefficients				
r	s	$^0\chi$	$^1\chi$	$^1\chi^v$	$^3\chi_P^v$	Constant
0.9807	4.236	14.47				−2.765
0.9819	4.108	—	22.88			4.211
0.9949	2.266	—	−29.71	51.56		25.74
0.9992	0.896	—	—	22.82	7.619	10.90
0.9993	0.853	1.296	—	20.87	6.496	9.59

TABLE XXIX

Observed and Calculated Values for Molar Magnetic Susceptibility χ_M *for Aliphatic Alcohols Correlated with the Connectivity Function*

		χ_M	
Compound		Obs	Calc
1.	Methanol	21.40	21.57
2.	Ethanol	33.60	33.15
3.	n-Propanol	45.18	45.80
4.	2-Propanol	45.79	45.39
5.	n-Butanol	56.54	57.15
6.	2-Methylpropanol	57.70	57.01
7.	2-Butanol	57.68	57.05
8.	2-Methyl-2-propanol	57.42	59.00
9.	n-Pentanol	67.50	68.50
10.	2-Methylbutanol	68.96	68.36
11.	2-Pentanol	69.10	68.40
12.	2-Methyl-2-butanol	70.90	69.80
13.	n-Hexanol	79.20	79.85
14.	4-Heptanol	91.50	91.54
15.	n-Octanol	102.65	102.55

error around the 1% level. Observed and predicted values are shown in Table XXIX.

V. Summary

In this chapter we have found significant correlations between the connectivity function and diverse properties of isolated molecules. The three principal properties examined here relate to the stability or total energy of the molecule, molecular volume, and the polarizability of the electron distribution. The various quantities discussed and correlated each relate to one or more of these basic properties.

It is especially noteworthy that the connectivity function correlations appear best where good data are available in sufficient quantity to permit meaningful multiple correlation.* Thermochemical and molar refraction

---

* We hold that there must be at least five observations for each independent variable in order that the correlation have statistical significance.

data for hydrocarbons and alcohols yield correlations with standard errors close to the estimated experimental error.

Traditionally, an understanding of the basic molecular properties examined here is considered fundamental to an understanding of the broader reaches of chemical science. As more chemical systems are examined with the connectivity function, the significance of various topological characteristics may become more apparent. It is hoped that the topological insights derived from molecular connectivity will contribute to further understanding of problems now facing the chemist and biologist.

Significant areas of the chemical and biological sciences deal not with molecules in isolation but with the interactions of molecules. With the development of methods for treating the isolated molecule presented here, we move in Chapter 5 to properties arising from molecular interactions.

References

1. J. D. Cox and G. Pilcher, "Thermochemistry of Organic and Organometalic Compounds." Academic Press, New York, 1970.
2. L. H. Hall, L. B. Kier, and W. J. Murray, unpublished work.
3. (a) G. R. Somayajulu and B. J. Zwolinski, *Trans. Faraday Soc.* **62,** 2327 (1966).
   (b) G. R. Somayajulu and B. J. Zwolinski, *Trans. Faraday Soc.* **68,** 1971 (1972).
4. (a) F. E. Norrington, R. M. Hyde, S. G. Williams, and R. Wooten, *J. Med. Chem.* **18,** 604 (1975).
   (b) C. Hansch, A. Leo, S. Unger, K. Kim, D. Nikaitani, and E. Lien, *J. Med. Chem.* **16,** 1209 (1973).
5. L. B. Kier, L. H. Hall, W. J. Murray, and M. Randić, *J. Pharm. Sci.* **64,** 1971 (1975).
6. L. B. Kier and L. H. Hall, *J. Pharm. Sci.* (in press).
7. J. A. Ostrenga, *J. Med. Chem.* **12,** 349 (1969).
8. L. J. Mullins, *Chem. Rev.* **54,** 289 (1954).
9. R. J. Wulf and R. M. Featherstone, *Anesthesiology* **18,** 97 (1957).
10. A. Cammarata, S. J. Yau, and K. S. Rogers, *J. Med. Chem.* **14,** 1211 (1971).
11. J. A. Beattie and O. C. Bridgeman, *Phys. Rev.* **35,** 643 (1930).
12. "Reference Data for Hydrocarbons and Petro-sulfur Compounds." Phillips Petroleum Co., Bartlesville, Oklahoma, 1962.

Chapter Five

MOLAR PROPERTIES AND MOLECULAR CONNECTIVITY

Physicochemical properties, broadly classified as molar or bulk, are functions of the intrinsic structure of each molecule, in addition to the interactions with neighboring molecules. The magnitude of these physical properties is a complex summation of internal and external forces in various combinations depending on the property.

Obviously, the intermolecular interactions depend on structural characteristics. It is the intrinsic structure, for example, that influences the intermolecular fit and optimum approach of large portions of neighboring molecules. This in turn may strengthen attractions that raise the values of melting point, heat of vaporization, etc.

For example, a more spherical or globular molecule such as neopentane possesses weaker intermolecular forces (lower boiling point and heat of vaporization) than a more linear or extended molecule such as *n*-pentane. These differences in intermolecular interaction may be traced to differences in the intrinsic nature of molecular structure or to the molecular topology.

The nature of the molecular structure also influences such geometric properties as surface area and molecular volume. These factors play an important role in such bulk properties as liquid density and water solubility. Further, molecular structure determines the electron distribution and its polarizability. Both these basic electron properties are largely responsible for intermolecular forces in addition to such factors as dipole moment and hydrogen bonding.

Bulk properties result from the complex interplay of several effects. Each of these effects may be described in terms of molecular structure. It is expected that most factors may be adequately related to the

TABLE I

Illustration of Influence of Structural Modification on Properties

Compound	Molecular weight	Boiling point	Density (D_4^{20})
Butane	58.13	−0.50	0.5788
cis-Butene-2	56.12	3.72	0.6306
Methylethyl ether	60.10	10.8	0.7260
Acetone	58.08	56.0	0.7899
1-Propanol	60.11	97.1	1.3850
Acetic acid	60.05	118.2	1.0493

molecular connectivity terms but some factors may require additional descriptors.

I. Intermolecular Forces Influenced by Intrinsic Structure

Intermolecular interaction has been analyzed and categorized into several types of forces such as (London) dispersion, polarization (including dipolar and induced dipolar forces), electrostatic, and repulsion arising from electron cloud overlap. A separate category is often reserved for hydrogen bonding. These forces vary in magnitude and relative importance with the degree of polarity and the size of the molecule.

In alkanes only dispersion forces are significant. The energy of interaction for alkanes of 5 to 10 carbon atoms falls in the range 0.5–1.0 kcal/mole. When a heteroatom such as nitrogen or oxygen is introduced, the resulting molecular polarity brings both polarization and electrostatic effects into play. Typical interaction energies for ethers and ketones are 1–5 kcal. In neutral organic molecules, even when charges are as high as ±0.3 electron in carbonyl groups, electrostatic effects are insignificant. Carbonyl and C–N groups give rise to large polarizations and dipolar forces. In the crystalline phase, large electrostatic interactions may occur between the charged organic species and their gegenions.

The increasing role of higher attractive interactions due to structural changes can be illustrated by examining a property such as the boiling point (Table I). Using a roughly constant molecular weight in this series, we introduce atoms or groups capable of stronger intermolecular interactions. As a result, the increased intermolecular attraction leads to increasing boiling point in the series.

We can analyze approximately the forces involved within each substance in the bulk. The hydrocarbon butane would associate with neighboring molecules almost exclusively through dispersion forces. The weak total interaction results in the lowest boiling point in the series. The introduction of a double bond, as in *cis*-butene, results in the introduction of some polarizability interaction, leading to a higher boiling point. In the ether, the oxygen atom introduces in the molecule a highly polar center, which is capable of dipole–dipole interactions with neighboring molecules, resulting in higher attractive forces. The boiling point correspondingly is measured higher.

In acetone the presence of the electronegative oxygen atom, in addition to a double bond, results in a combination of dipolar and polarization forces, which raises the intermolecular attractive force still higher. The boiling point rises significantly over the ether of approximately the same molecular weight and dispersion properties. Propanol possesses the capability of strong intermolecular hydrogen bonding in addition to the dipole forces derived from the carbon–oxygen bond. This leads to the higher boiling point observed in this case.

Finally, acetic acid has two oxygen atoms with their attendant dipolar forces, polarizable double bond, and the capability of donating and receiving hydrogen bonds within the same molecule. This results in the strong intermolecular attractions found in acids, leading to the highest boiling points in the series.

As we have seen in the preceding chapter, topology in the form of molecular connectivity can be useful in describing structural characteristics influencing molecular properties. To the extent that molar properties are related to structural characteristics represented by molecular connectivity, we can expect that expressions of χ should be sufficiently descriptive of structure so as to be useful in correlating and predicting physicochemical properties.

Before discussing some of these relationships, it should be pointed out that success in relating functions of χ to molar properties is but a step away from the successful use of molecular connectivity to study structure–activity relationships among drug molecules.

II. Heat of Vaporization

The heat required to vaporize a substance in the liquid phase is called the heat of vaporization. If q_p is the heat (kcal/mole) required to vaporize one mole of liquid at constant temperature and pressure, the following relations to enthalpy and entropy hold:

$$q_p = \Delta H_{vap} = H_{vap} - H_{liq}$$
$$\Delta S = S_{vap} - S_{liq} = \Delta H_{vap}/T_b$$

where T_b is the normal boiling point. Heats of vaporization are usually tabulated for the normal boiling point at 1 atm pressure.

The energy of vaporization ΔE_{vap} is related to ΔH_{vap} by the standard thermodynamic relation

$$\Delta E_{vap} = \Delta H_{vap} - RT_b$$

Thus ΔH_{vap} is a direct measure of the magnitude of the intermolecular forces such as we have just described.

We would anticipate that molecular connectivity describes some of the important structural characteristics contributing to the magnitude of ΔE_{vap}. We also expect that structural definitions based on molecular connectivity would be applicable to series of closely related molecules. Major differences in electronic structure as found in different classes of organic molecules introduce features that are not described quantitatively at this time. Applied quantum mechanical calculations may be required to supplement the connectivity approach.

A. Alkanes

Excellent data are available for 44 alkanes, which range from ethane to decane and which include a variety of structural patterns. Somayajulu and Zwolinski have estimated the probable experimental standard deviation as 0.030 kcal/mole [1].

Correlation with just the number of carbon atoms n leads to a standard error of 0.476 kcal. By contrast, the standard error is significantly lower at 0.163 by correlation with $^1\chi$ [2]:

$$\Delta H_{vap} = 0.486 + 2.381\ ^1\chi$$

In their study of this set of data, Somayajulu and Zwolinski used a 13-variable equation, based on various structural considerations, to obtain a standard error of 0.046 kcal/mole [1]. We have improved upon this correlation appreciably by using only seven $^m\chi_t$ terms, obtaining a standard error of 0.042 kcal/mole. The seven-variable function used by

TABLE II

Statistical Data for Alkane Heat of Vaporization ΔH_{vap} Correlated with the Connectivity Function

Statistical data			Regression coefficients								
r	s	n	$^1\chi$	$^2\chi$	$^4\chi_P$	$^5\chi_P$	$^6\chi_P$	$^3\chi_C$	$^4\chi_C$	$^6\chi_{PC}$	Constant
0.9834	0.476	1.165									-0.0432
0.9981	0.163	—	2.381								0.4860
0.9981	0.164	-0.063	2.506								0.5268
0.9999	0.042	—	1.752	1.301	-0.4150	-0.4366	-0.2659	-2.091	3.051		0.5673
0.9999	0.042	—	1.774	1.265	-0.3922	-0.4069	-0.3323	-2.016	2.925	-0.0039	0.5470

TABLE III

Observed and Calculated Values for Alkane Heat of Vaporization ΔH_{vap}
Correlated with the Connectivity Function

| | | ΔH_{vap} | |
Compound		Obs	Calc
1.	Ethane	2.264	2.319
2.	Propane	3.965	3.965
3.	n-Butane	5.191	5.222
4.	Isobutane	4.799	4.647
5.	n-Pentane	6.395	6.411
6.	2-Methylbutane	6.030	6.035
7.	2,2-Dimethylpropane	5.345	5.317
8.	n-Hexane	7.555	7.577
9.	2-Methylpentane	7.160	7.166
10.	3-Methylpentane	7.255	7.263
11.	2,2-Dimethylbutane	6.651	6.659
12.	2,3-Dimethylbutane	6.985	7.040
13.	n-Heptane	8.739	8.747
14.	2-Methylhexane	8.325	8.309
15.	3-Methylhexane	8.391	8.375
16.	3-Ethylpentane	8.425	8.363
17.	2,2-Dimethylpentane	7.764	7.740
18.	2,3-Dimethylpentane	8.191	8.175
19.	2,4-Dimethylpentane	7.872	7.878
20.	3,3-Dimethylpentane	7.901	7.905
21.	2,2,3-Trimethylbutane	7.669	7.730
22.	n-Octane	9.915	9.935
23.	2-Methylheptane	9.484	9.478
24.	3-Methylheptane	9.521	9.522
25.	4-Methylheptane	9.483	9.491
26.	3-Ethylhexane	9.476	9.456
27.	2,2-Dimethylhexane	8.915	8.862
28.	2,3-Dimethylhexane	9.272	9.268
29.	2,4-Dimethylhexane	9.029	9.068
30.	2,5-Dimethylhexane	9.051	9.022
31.	3,3-Dimethylhexane	8.973	8.969
32.	3,4-Dimethylhexane	9.316	9.294
33.	2-Methyl-3-ethylpentane	9.209	9.194
34.	3-Methyl-3-ethylpentane	9.081	9.043
35.	2,2,3-Trimethylpentane	8.826	8.790
36.	2,2,4-Trimethylpentane	8.402	8.415
37.	2,3,3-Trimethylpentane	8.897	8.909
38.	2,3,4-Trimethylpentane	9.014	9.030
39.	2,2,3,3-Tetramethylbutane	8.410	8.412

TABLE III—Continued

| Compound | ΔH_{vap} | |
	Obs	Calc
40. *n*-Nonane	11.10	11.12
41. 2,2,3,3-Tetramethylpentane	9.871	9.867
42. 2,2,3,4-Tetramethylpentane	9.478	9.589
43. 2,2,4,4-Tetramethylpentane	9.580	9.557
44. 2,3,3,4-Tetramethylpentane	9.910	9.945
45. *n*-Decane	12.276	12.304
46. *n*-Dodecane	14.650	14.673
47. *n*-Hexadecane	19.450	19.410

Somayajulu and Zwolinski developed a considerably larger error, 0.062 [1]. The connectivity function is

$$\Delta H_{vap} = 0.567 + 1.752\ ^1\chi + 1.301\ ^2\chi - 0.415\ ^4\chi_P - 0.437\ ^5\chi_P$$
$$- 0.266\ ^6\chi_P - 2.091\ ^3\chi_C + 3.051\ ^4\chi_C$$
$$r = 0.9999, \quad s = 0.042, \quad N = 44$$

Statistical data are summarized in Table II. Observed and predicted values of ΔH_{vap} are shown in Table III.

We can consider this to be a remarkable achievement. With this number of variables, the connectivity function predicts a property in the bulk phase. Values for $^m\chi_t$ terms derive from graphs for isolated molecules. These terms must be significant for the description of structure as it relates to intermolecular forces.

It would be desirable to analyze each coefficient and $^m\chi_t$ variable in terms of intermolecular forces. It is hoped that this will be done in the future so that coefficients may be calculated from a model of a property, rather than by regression analysis. Then connectivity methods could be extended to compounds and properties for which experimental data are now too limited for meaningful multiple regression.

B. Alcohols

Less data are available for alcohols than for hydrocarbons, but the following analysis has been performed for this important class of

TABLE IV

Observed and Calculated Values for Heat of Vaporization ΔH_{vap} of Aliphatic Alcohols Correlated with the Connectivity Function

			ΔH_{vap}	
Compound		Obs	Calc[a]	Calc[b]
1.	Methanol	8.94	8.96	8.84
2.	Ethanol	10.18	10.13	9.99
3.	n-Propanol	11.34	11.29	11.47
4.	2-Propanol	10.90	10.54	11.00
5.	n-Butanol	12.50	12.45	12.65
6.	2-Methylpropanol	12.15	11.86	12.08
7.	2-Butanol	11.89	11.86	11.60
8.	2-Methyl-2-propanol	11.14	10.76	11.14
9.	n-Pentanol	13.61	13.61	13.77
10.	2-Pentanol	12.56	13.02	12.64
11.	3-Pentanol	12.36	13.18	12.42
12.	2-Methyl-1-butanol	13.04	13.18	13.16
13.	3-Methyl-1-butanol	13.15	13.02	13.09
14.	3-Methyl-2-butanol	12.27	12.51	12.22
15.	n-Hexanol	15.00	14.78	14.93
16.	n-Heptanol	16.20	15.94	16.09
17.	n-Octanol	17.00	17.10	17.24
18.	2-Ethyl-1-hexanol	16.12	16.82	16.45
19.	n-Nonanol	18.60	18.27	18.40
20.	n-Decanol	19.82	19.43	19.55

[a] Based on Eq. (1).   [b] Based on Eq. (2).

compounds. Correlation with n is unsatisfactory, but the addition of $^1\chi$ yields the equation [2]

$$\Delta H^i_{vap}(n) = 6.641 - 4.083\,\Delta\chi_i + 1.163n \tag{1}$$

$$r = 0.9926, \qquad s = 0.362, \qquad N = 20$$

The corresponding equation for hydrocarbons reveals a useful comparison:

$$\Delta H^i_{vap}(n) = 0.444 - 2.404\,\Delta\chi_i + 1.171n$$

where $\Delta\chi_i = \chi_n - {}^1\chi_i$.

The coefficient of $\Delta\chi_i$ for alcohols is almost twice that for hydrocarbons, indicating that the effect of branching in alcohols is more significant than in hydrocarbons. On the other hand, the coefficient of the number of nonhydrogen atoms is the same, within experimental error, for both sets of compounds. For monoalcohols this number n is simply the count of methylene groups. This suggests that the contribution of each carbon is the same in both sets of compounds. The constant term is about 6 kcal/mole larger for alcohols, reflecting the difference in intermolecular forces. The magnitude of hydrogen bonding effects in alcohols is considered to be about this same value. Thus, even at this level, the connectivity method appears to give a reasonable basis for describing the effects at work in the vaporization process. Observed and predicted values for alcohols are shown in Table IV.

The connectivity function based on the valence δ may be used for alcohols. When the $^1\chi^v$ term is added to the $^1\chi$ term, the correlation is of the same quality as above:

$$\Delta H_{vap} = 6.593 - 1.063\ {}^1\chi^v + 4.901\ {}^1\chi - 0.7614n$$

$$r = 0.9927, \qquad s = 0.371, \qquad N = 20$$

When additional higher order extended terms are used in the connectivity function, the correlation is improved. However, because of the limited set of data, a thorough investigation is not possible at this time. The standard error is about 50% less in a six-variable function:

$$\Delta H_{vap} = 8.491 + 9.111\ {}^1\chi - 8.764\chi_n - 2.416\ {}^3\chi_C$$

$$+ 3.206\ {}^2\chi^v - 0.856\ {}^4\chi_P^v + 6.506\ {}^3\chi_C^v \qquad (2)$$

$$r = 0.9982, \qquad s = 0.204, \qquad N = 20$$

The predicted values of ΔH_{vap} are listed in the right-hand column of Table IV. It should be pointed out that there are no outstandingly large errors. Thus, the connectivity function based on valence δ values appears to give a good account of heat of vaporization. This property depends chiefly on intermolecular forces, as does boiling point.

III. Boiling Point

The normal boiling point T_b of a liquid is defined as the temperature at which the vapor pressure reaches 1 atm. Boiling points may be measured directly or, in a more accurate approach, extrapolated from vapor pressure data.

For nonpolar molecules the boiling point is dependent on polarizability through the effect of London dispersion forces. For polar molecules, the effects attendant to a permanent dipole increase the magnitude of the

TABLE V

Observed and Calculated Values for Alkane Boiling Point Correlated with the Connectivity Function

		Boiling point	
	Compound	Obs	Calc
1.	2,2-Dimethylpropane	9.5	15.2
2.	2-Methylbutane	27.9	32.2
3.	*n*-Pentane	36.1	38.3
4.	2,2-Dimethylbutane	49.7	51.4
5.	2,3-Dimethylbutane	57.9	57.3
6.	2-Methylpentane	60.2	59.4
7.	3-Methylpentane	63.5	62.9
8.	*n*-Hexane	68.7	66.1
9.	2,2,3-Trimethylbutane	80.9	80.0
10.	2,2-Dimethylpentane	79.2	77.8
11.	3,3-Dimethylpentane	86.0	85.7
12.	2,4-Dimethylpentane	80.5	80.1
13.	2,3-Dimethylpentane	89.7	87.0
14.	2,2,3,3-Tetramethylbutane	106.3	106.0
15.	2-Methylhexane	90.0	87.3
16.	3-Methylhexane	93.4	93.0
17.	3-Ethylpentane	93.4	93.0
18.	*n*-Heptane	99.3	97.9
19.	2,2,4-Trimethylpentane	99.3	97.9
20.	2,2,3-Trimethylpentane	109.9	108.5
21.	2,3,3-Trimethylpentane	114.8	112.8
22.	2,3,4-Trimethylpentane	113.5	110.3
23.	2,2-Dimethylhexane	106.9	105.6
24.	3,3-Dimethylhexane	112.0	112.4
25.	2,5-Dimethylhexane	109.1	108.5
26.	2,4-Dimethylhexane	109.5	111.1
27.	2,3-Dimethylhexane	115.6	114.5
28.	3,3-Methylethylpentane	118.3	118.4
29.	2,2,4,4-Tetramethylpentane	122.3	115.3
30.	3,4-Dimethylhexane	117.7	116.9
31.	2,3-Methylethylpentane	115.7	116.3
32.	2-Methylheptane	117.7	115.1
33.	3-Methylheptane	119.0	118.2
34.	4-Methylheptane	117.7	117.8
35.	2,2,3,3-Tetramethylpentane	140.3	137.4
36.	3-Ethylhexane	118.6	120.6
37.	2,2,3,4-Tetramethylpentane	133.1	131.2
38.	*n*-Octane	125.7	121.8
39.	2,4,4-Trimethylhexane	130.4	132.6
40.	2,2,5-Trimethylhexane	124.1	126.8

TABLE V—Continued

| | Boiling point | |
Compound	Obs	Calc
41. 2,2,4-Trimethylhexane	126.5	129.0
42. 2,2,3-Trimethylhexane	133.6	136.1
43. 2,3,3-Trimethylhexane	137.7	139.6
44. 2,2-Dimethyl-3-ethylpentane	133.8	136.8
45. 2,3,5-Trimethylhexane	131.3	135.3
46. 3,3,4-Trimethylhexane	140.5	141.7
47. 2,2-Dimethylheptane	132.7	133.5
48. 2,4-Dimethyl-3-ethylpentane	136.7	138.9
49. 3,3-Diethylpentane	146.2	150.1
50. 4-Methyloctane	142.5	145.7
51. n-Nonane	150.8	149.6

boiling temperature. The boiling point is a classical value used in organic qualitative analysis for the identification of a molecule. It is a characteristic property that can be evaluated easily.

A relation between boiling point and intermolecular forces is revealed in the following equation:

$$\Delta E_{vap} = \Delta H_{vap} - RT_b$$

where E is the internal energy, H the enthalpy, and R the molar gas constant, -1.987 cal/deg-mole. Since entropy of vaporization is related to enthalpy as

$$\Delta S_{vap} = \Delta H_{vap}/T_b$$

then

$$\Delta E_{vap} = T_b(\Delta S_{vap} - R)$$

For liquids such as typical organic compounds, whose vapors follow a nearly ideal behavior and obey Trouton's rule, the entropy of vaporization is roughly a constant:

$$\Delta S_{vap} \approx 21 \quad \text{cal/deg-mole}$$

and ΔH_{vap} is approximately 21 times the boiling point T_b. Thus there is an approximate correspondence of magnitude of boiling temperature to the magnitude of intermolecular attractions:

$$\Delta E_{vap} \approx 19T_b$$

The boiling point is then a measure of the intermolecular forces binding molecules in the liquid state.

By a line of reasoning similar to that developed in the section on heat of vaporization, we would expect that terms in the connectivity function would mirror enough of the structure to be able to give significant correlation with the boiling point.

A. Alkanes

In the second paper published on molecular connectivity [3] a correlation was developed between $^1\chi$ and the boiling point of saturated hydrocarbons. Both cyclics and noncyclics were included. For a list of 62 compounds a standard error of 8.5° is obtained.

For the present study only noncyclic compounds are considered for which good data are available. For the list of 51 molecules shown in Table V, correlation with $^1\chi$ is quite good:

$$\text{b.p.} = 57.85 \, ^1\chi - 97.90, \qquad r = 0.9851, \quad s = 5.51, \quad N = 51$$

Significant improvement is obtained by the addition of one subgraph term, $^4\chi_{PC}$:

$$\text{b.p.} = 55.69 \, ^1\chi + 4.708 \, ^4\chi_{PC} - 96.13$$

$$r = 0.9969, \qquad s = 2.53, \qquad N = 51$$

Results predicted by this equation are given in Table V.

B. Alcohols

Also in the second paper on molecular connectivity [3] the solubility and boiling points of alcohols were investigated. Correlation with $^1\chi$ alone yields a standard error of about 8°. Part of the problem in this study is the occurrence of redundancies, compounds differing in OH group position with the same $^1\chi$ value but different boiling points. The compounds 2-methylpropanol and *sec*-butanol are examples of this redundancy.

Hall *et al.* [3] have introduced an empirical modification to take account of the problem of redundancies. A second regression variable is introduced. The additional variable is the connectivity contribution for the OH bond, c_{OH}. Since this term includes the δ value for the attached carbon atom, c_{OH} varies with position of substitution: 0.707 for primary, 0.577 for secondary, and 0.500 for tertiary. Hence, the use of c_{OH} destroys the redundancies and improves the correlation significantly.

For a set of 62 alcohols including 11 monocyclic alcohols, the

TABLE VI

Regression Statistical Analysis for Correlation of Aliphatic Alcohol Boiling Point with the Connectivity Function

Statistical data		Regression coefficients					
r	s	$^1\chi$	$^1\chi^v$	$^3\chi_P$	$^3\chi_P^v$	$^3\chi_C$	Constant
0.9459	9.33	34.93					25.90
0.9943	3.12	212.16	−174.33				−43.33
0.9976	2.05	203.05	−161.58	−6.21			−41.96
0.9991	1.28	212.66	−175.41	−17.95	20.07		−38.64
0.9996	0.86	191.83	−155.23	−17.99	21.37	−2.74	−29.05

regression equation and statistics are

$$\text{b.p.} = -35.29 + 37.56\,{}^1\chi + 89.00c_{\text{OH}}, \qquad r = 0.980, \quad s = 6.75, \quad N = 62$$

The corresponding statistical information for the regression with $^1\chi$ alone is

$$\text{b.p.} = 20.29 + 36.85\,{}^1\chi, \qquad r = 0.958, \quad s = 9.66, \quad N = 62$$

In order to provide a basis for a meaningful study with the connectivity function, careful attention must be given to the reliability and quality of the data. For example, the boiling point of a *dl* mixture is known to be lower than that for a pure compound. Also many boiling points are recorded as a 1 or 2° range. When these are purged from the list, there remain 28 compounds for which there are reliable data.

When the connectivity function is applied to the list of aliphatic alcohols, a remarkably good correlation results. The regression data are summarized in Table VI, and the best equation is

$$\text{b.p.} = 191.83\,{}^1\chi - 155.23\,{}^1\chi^v - 17.99\,{}^3\chi_P$$

$$+ 21.74\,{}^3\chi_P^v - 2.74\,{}^3\chi_C - 29.05$$

$$r = 0.9996, \qquad s = 0.86, \qquad N = 28$$

A relative error of 0.6% must be considered excellent. It should be noted that the higher order terms are all of order three. Both the valence and connectivity path terms as well as the connectivity cluster term are found to be significant. Values calculated from parameters in the last line of Table VI for the boiling point are found in Table VII.

TABLE VII

Observed and Calculated Values for Aliphatic Alcohol Boiling Point Correlated with the Connectivity Function

		Boiling point	
Compound		Obs	Calc
1.	2-Propanol	82.4	82.3
2.	1-Propanol	97.1	97.5
3.	1-Butanol	117.6	118.3
4.	2-Methyl-propanol	108.1	106.8
5.	2-Methyl-2-butanol	102.3	103.0
6.	2-Pentanol	118.9	120.5
7.	3-Methyl-1-butanol	132.0	131.6
8.	2-Methyl-1-butanol	128.9	130.4
9.	1-Pentanol	138.0	137.6
10.	3,3-Dimethyl-2-butanol	120.4	119.7
11.	2-Methyl-2-pentanol	123.0	122.3
12.	3-Methyl-3-pentanol	136.0	123.4
13.	2,2-Dimethyl-1-butanol	136.7	138.2
14.	4-Methyl-2-pentanol	133.0	133.3
15.	2-Methyl-3-pentanol	127.5	129.1
16.	2,3-Dimethyl-1-butanol	145.0	145.4
17.	2-Methyl-1-pentanol	148.0	149.0
18.	1-Hexanol	157.5	156.8
19.	2,4-Dimethyl-3-pentanol	140.0	140.0
20.	2,4-Dimethyl-1-pentanol	159.8	161.3
21.	3-Ethyl-3-pentanol	142.0	142.9
22.	1-Heptanol	176.8	176.1
23.	4-Methyl-4-heptanol	161.0	162.3
24.	3-Methyl-3-heptanol	163.0	162.2
25.	6-Methyl-1-heptanol	188.0	188.6
26.	1-Octanol	194.4	195.3
27.	2-Methyl-2-octanol	178.0	180.1
28.	4-Ethyl-4-heptanol	182.0	182.0

C.   Aliphatic Ethers

There is a very limited set of reliable data for ethers. For the 12 compounds shown in Table VIII the correlation with $^1\chi$ yields $r = 0.9851$ and $s = 5.69$. The addition of $^1\chi^v$ gives only slight improvement:

$$\text{b.p.} = 79.15\ {}^1\chi - 30.90\ {}^1\chi^v - 83.75$$

$$r = 0.9882, \qquad s = 5.39, \qquad N = 11$$

Observed and calculated results are given in Table VIII. There are insufficient data to warrant addition of extended terms, which might provide an adequate description of such diverse structural cases as methyl-*tert*-butyl ether and diisopropyl ether.

D. Aliphatic Amines

There are three classes of amines corresponding to the degree of substitution on the nitrogen atom. These classes, primary (RNH_2), secondary (R_2NH), and tertiary (R_3N), have greatly differing boiling points. The number of hydrogen atoms on the nitrogen available for hydrogen bonding greatly influences the strength of the intermolecular forces. Also, the electron density on the nitrogen atom is influenced by the degree of substitution. For these reasons the three classes must be considered separately. These effects are illustrated by the fact that *n*-butyl amine boils 25° higher than diethyl amine. Both compounds have the same $^1\chi$, 2.914.

Good data are available for 12 primary amines. Correlations with $^1\chi$ and $^1\chi^v$ are summarized as follows:

$$\text{b.p.} = 54.50 \;^1\chi - 58.07, \qquad r = 0.9936, \quad s = 4.00, \quad N = 12$$

$$\text{b.p.} = 151.74 \;^1\chi - 98.33 \;^1\chi^v - 81.59$$

$$r = 0.9991, \qquad s = 1.56, \qquad N = 12$$

TABLE VIII

Observed and Calculated Values for Ether Boiling Point Correlated with the Connectivity Function

		Boiling point	
Compound		Obs	Calc
1.	Methyl propyl ether	38.9	43.6
2.	Methyl *tert*-butyl ether	55.2	53.6
3.	Ethyl *tert*-butyl ether	73.1	75.0
4.	*sec*-Butyl ethyl ether	81.0	84.0
5.	*sec*-Butyl methyl-2-methyl ether	86.3	76.4
6.	Ethyl *tert*-pentyl ether	101.0	101.0
7.	Butyl isopropyl ether	108.0	106.3
8.	Hexyl methyl ether	118.0	118.1
9.	Ethyl pentyl ether	123.0	120.8
10.	Di-*sec*-butyl ether	121.0	129.6
11.	Dibutyl ether	142.0	138.2

For secondary amines

$$\text{b.p.} = 48.50 \, {}^1\chi - 58.55, \qquad r = 0.997, \quad s = 4.28, \quad N = 13$$

$$\text{b.p.} = 171.40 \, {}^1\chi - 120.07 \, {}^1\chi^v - 102.4$$

$$r = 0.999, \qquad s = 2.20, \qquad N = 13$$

In both cases the improvement in the correlation is dramatic upon the addition of the valence-dependent variable ${}^1\chi^v$.

There are only four tertiary amines with well-determined boiling points. The correlation with ${}^1\chi$ or ${}^1\chi^v$ leads to $r \approx 0.97$, but the best correlation is found with ${}^0\chi^v$:

$$\text{b.p.} = 31.83 \, {}^0\chi^v - 67.18, \qquad r = 0.997, \quad s = 2.92, \quad N = 4$$

Some of the differences between the three classes of amines may be

TABLE IX

Observed and Calculated Values for Amine Boiling Points Correlated with the Connectivity Function

		Boiling point	
Compound		Obs	Calc
1.	Propyl amine	49.0	52.4
2.	2-Amino-2-methylpropane	44.4	45.1
3.	Isopropylmethyl amine	50.0	55.4
4.	Diethyl amine	56.0	58.3
5.	1-Aminobutane	77.8	77.9
6.	2-Amino-2-methylbutane	78.0	73.7
7.	1-Amino-3-methylbutane	95.0	96.1
8.	Isopentyl amine	96.0	96.1
9.	3-Aminopentane	91.0	90.7
10.	1-Aminopentane	104.4	103.5
11.	3-Amino-2,2-dimethylbutane	102.0	97.6
12.	Di-isopropyl amine	84.0	87.4
13.	Butyldimethyl amine	95.0	91.0
14.	Triethyl amine	89.0	84.7
15.	Hexyl amine	130.0	129.0
16.	Dimethylpentyl amine	123.0	116.6
17.	2-Aminoheptane	142.0	139.8
18.	1-Aminoheptane	156.9	154.5
19.	Di-isobutyl amine	139.0	145.7
20.	Dibutyl amine	159.0	160.4
21.	Tripropyl amine	156.0	161.3

TABLE X

Observed and Calculated Values for Boiling Point of Polyols Correlated with Connectivity

	Boiling point	
Compound	Obs	Calc
1. 2-Methyl-2,4-pentanediol	196.0	197.3
2. Ethanediol	197.8	197.3
3. 1,3-Butanediol	207.5	204.8
4. 1,3-Propanediol	214.7	211.0
5. 1,2-Propanediol	187.4	191.1
6. 2,3-Butanediol	181.7	185.6
7. 1,4-Butanediol	230.0	224.8
8. 1,5-Pentanediol	238	238.5
9. Glycerol	290	292.7

taken into account by the addition of a variable that describes the local connectivity at the nitrogen. The valence δ_N^v is such a variable. δ_N^v also is linearly related to the number of hydrogen atoms on the nitrogen:

$$\delta_N^v = 5 - h$$

When the three classes of amines are taken together, correlation with $^1\chi$, $^1\chi^v$, and δ_N^v gives the following:

$$\text{b.p.} = 184.58 \, {}^1\chi - 133.51 \, {}^1\chi^v - 18.87\delta_N^v - 28.57$$

$$r = 0.9949, \qquad s = 3.94, \qquad N = 21$$

Observed and calculated values for amine boiling points are given in Table IX.

The addition of other extended terms does decrease the standard error. The use of $^3\chi_P^v$ and $^3\chi_P$ decreases s to 2.34. However, because there are only 21 observations, the statistical significance of these results is doubtful. At the present time we will limit our work with amines to three variables.

E. Aliphatic Polyols

A dramatic illustration of the usefulness of the valence connectivity terms is found in boiling point data for diols. Table X lists eight dihydroxy and one trihydroxy compound. As one might expect, very

TABLE XI

Observed and Calculated Values for Alkyl Halide Boiling Point Correlated with the Connectivity Function

Compound	Boiling point	
	Obs	Calc
1. 2-Chloropropane	36.5	38.2
2. *tert*-Butyl chloride	51.0	52.0
3. 2-Chlorobutane	68.0	70.5
4. 1-Chloro-2-methylpropane	69.0	71.2
5. 2-Chlorobutane	78.0	73.5
6. 3-Chloropentane	97.0	99.1
7. 2-Bromopropane	60.0	56.2
8. 1-Bromopropane	71.0	72.4
9. 1-Bromo-2-methylpropane	91.0	91.7
10. 2-Bromobutane	91.0	88.7
11. 3-Bromopentane	118.0	116.5
12. 2-Iodopropane	90.0	91.0
13. 2-Iodobutane	120.0	122.2
14. 1-Iodobutane	130.0	133.4
15. 2-Iodopentane	142.0	141.2
16. 1-Iodopentane	155.0	155.7
17. 1-Iodohexane	179.0	178.0

poor correlation is found with $^1\chi$:

$$\text{b.p.} = 20.71\ ^1\chi + 159.38, \quad r = 0.307, \quad s = 34.0, \quad N = 9$$

The addition of the $^1\chi^v$ term dramatically cuts the standard error tenfold:

$$\text{b.p.} = 249.64\ ^1\chi - 222.1\ ^1\chi^v - 29.11, \quad r = 0.995, \quad s = 3.78, \quad N = 9$$

Calculated boiling points are found in Table X [4].

F. Alkyl Halides

For a list of 17 mixed chlorides, bromides, and iodides, correlation with $^1\chi$ is not expected to yield good correlation: $r = 0.839$. However, use of the connectivity function produces results comparable to those for alcohols and amines:

$$\text{b.p.} = 39.91 \; {}^1\chi^v - 29.12, \qquad r = 0.9877, \quad s = 6.25, \quad N = 17$$

$$\text{b.p.} = 36.26 \; {}^1\chi^v + 16.36 \; {}^3\chi_P + 2.869 \; {}^3\chi_C^v - 31.037$$

$$r = 0.9982, \qquad s = 2.57, \qquad N = 17$$

Observed and calculated values are given in Table XI.

This study appears to be a good test of the validity and usefulness of the δ^v values in that there are three types of compounds included, with various branching patterns.

IV.   Liquid Density

Density is defined as the ratio of mass to volume. It is a measure of the number of molecules that are contained in a unit volume. Density is determined by a geometric factor describing or relating to the size and shape of the electron cloud of the molecule. In addition, the magnitude of intermolecular forces determines how strongly or how loosely the

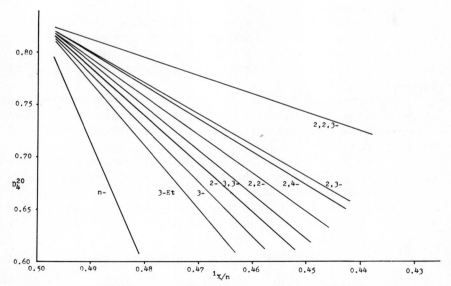

Fig. 1. A plot of liquid density D_4^{20} for alkanes. Density is plotted against ${}^1\chi/n$, where n is the number of carbon atoms in the alkane. Each line represents a structural class; for example, 2- stands for methyl substitution in the 2 position.

molecules are held together. The relative orientation of molecular species is also determined by intermolecular forces.

Since there is such a wide variety in the shapes of organic molecules as well as a dynamic interplay between the molecules under thermal motion, it is expected that the experimental value of density is a complex composite of several factors. As shown in Chapter 1, hydrocarbon density is not ranked properly by $^1\chi$ alone. On the other hand, properties of alkanes such as heat of atomization, boiling point, and heat of vaporization are ranked correctly by $^1\chi$.

A. Alkanes

An interesting analysis can be performed in terms of functions of $^1\chi$. For example, density plotted against $^1\chi/n$ produces a fan of curves (see Fig. 1). The $^1\chi/n$ value to which these lines converge is the maximum value that $^1\chi/n$ can achieve for large alkanes:

$$\lim_{n\to\infty} {}^1\chi/n = 0.5$$

The convergent density value appears to be the maximum value for alkane density, 0.83. The line with the steepest slope is that for straight-chain compounds, whereas the smallest slopes are for the highly branched compounds.

In a somewhat different analysis a set of parallel lines is obtained when n/d is plotted against $^1\chi$. The more branched compounds have the highest and lesser branched compounds the lowest value of intercept. Molar volume is directly related to n/d and is the basis for this family of curves.

These two plots are shown in Figs. 1 and 2. The slopes and intercepts of these sets of lines may be related to $^1\chi$ but the resulting relations between density and $^1\chi$ are quite cumbersome and not easily extended to heteroatomic systems.

For a set of 82 alkanes the correlation of density with $^1\chi$ alone gives a poor correlation [5]. From earlier analyses [6] it has appeared that $^1\chi$ bears a relationship to molecular volume. Hence, a more meaningful correlation to reciprocal $^1\chi$ might be expected. Such a regression yields $r = 0.903$. It appears that other extended terms are required for a quality fit. The best second variable is found to be $^3\chi_P$:

$$d = 0.7348 - 0.2929/{}^1\chi + 0.0030\ {}^3\chi_P$$

$$r = 0.9889, \qquad s = 0.0046, \qquad N = 82$$

This regression equation produces the correct ranking of densities in alkanes with the exception of two very closely spaced heptanes.

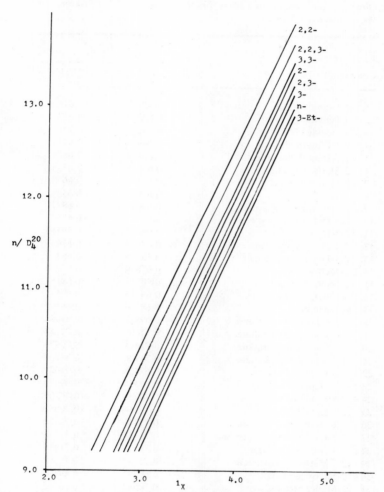

Fig. 2. A plot of n/D_4^{20} for alkanes, where D_4^{20} is the liquid density and n the number of carbon atoms in the alkane. The horizontal axis is $^1\chi$. Each line represents a different structural class; for example, 3-stands for methyl substitution at the 3 position.

The addition of other extended terms improves the relationship considerably. However, in order to provide a good test of the connectivity method for this property, we have developed a shorter list of alkanes for which the data are of the highest quality. The connectivity function has been tested against this list regardless of whether the compounds fit

TABLE XII

Observed and Calculated Values for Alkane Density D_4^{20} Correlated with the Connectivity Function

		D_4^{20}	
Compound		Obs	Calc
1.	2-Methylbutane	0.61967	0.61938
2.	*n*-Pentane	0.62624	0.62714
3.	*n*-Octane	0.70252	0.70125
4.	2,2-Dimethylbutane	0.64916	0.64858
5.	2,3-Dimethylbutane	0.66164	0.66751
6.	2-Methylpentane	0.65315	0.65452
7.	3-Methylpentane	0.66431	0.66500
8.	*n*-Hexane	0.65937	0.65892
9.	2,2,3-Trimethylbutane	0.69011	0.68777
10.	2,2-Dimethylpentane	0.67385	0.67582
11.	3,3-Dimethylpentane	0.69327	0.69540
12.	2,4-Dimethylpentane	0.67270	0.67637
13.	2,3-Dimethylpentane	0.69508	0.69301
14.	2-Methylhexane	0.67859	0.67820
15.	3-Methylhexane	0.68713	0.68757
16.	3-Ethylpentane	0.69816	0.69531
17.	*n*-Heptane	0.68376	0.68292
18.	2,2,4-Trimethylpentane	0.69192	0.69466
19.	2,2,3-Trimethylpentane	0.71602	0.71743
20.	2,3,3-Trimethylpentane	0.72619	0.72623
21.	2,3,4-Trimethylpentane	0.71906	0.71705
22.	2,2-Dimethylhexane	0.69528	0.69421
23.	2,5-Dimethylhexane	0.69354	0.69339
24.	2,4-Dimethylhexane	0.70036	0.70309
25.	2,3-Dimethylhexane	0.71214	0.70951
26.	3,3-Methylethylpentane	0.72742	0.72927
27.	2,2,4,4-Tetramethylpentane	0.71947	0.71150
28.	3,4-Dimethylhexane	0.71923	0.71896
29.	2,3-Methylethylpentane	0.71932	0.71665
30.	2-Methylheptane	0.69792	0.69867
31.	3-Methylheptane	0.70582	0.70562
32.	4-Methylheptane	0.70463	0.70510
33.	2,2,3,3-Tetramethylpentane	0.75666	0.75475
34.	3-Ethylhexane	0.71358	0.71008
35.	2,2,3,4-Tetramethylpentane	0.73895	0.73934
36.	2,3,3,4-Tetramethylpentane	0.75473	0.75377
37.	2,2,5-Trimethylhexane	0.70721	0.70664
38.	2,2,4-Trimethylhexane	0.71560	0.71708
39.	2,4,4-Trimethylhexane	0.72381	0.72620
40.	2,2,3-Trimethylhexane	0.72920	0.72985
41.	2,3,3-Trimethylhexane	0.73800	0.73850
42.	2,3,5-Trimethylhexane	0.72190	0.72180
43.	2,2-Dimethylheptane	0.71050	0.71255
44.	3,3-Diethylpentane	0.75359	0.75372
45.	4-Methyloctane	0.71990	0.71907
46.	*n*-Nonane	0.71763	0.71595

the above equation well or not. Only the quality of the data is used as a criterion and observations with five or six significant figures are used.

The use of a five-variable connectivity function yields a standard error about half that given above:

$$d = 0.7955 - 0.4288/^1\chi + 0.0131 \; ^3\chi_P$$

$$- 0.0079 \; ^5\chi_P + 0.0051 \; ^4\chi_{PC} + 0.0035 \; ^5\chi_{PC}$$

$$r = 0.9971, \quad s = 0.0024, \quad N = 46$$

Only one compound shows a residual greater than twice the standard error. Observed and calculated values are given in Table XII.

B. Aliphatic Alcohols

For 40 alcohols for which density values are expressed to four or five significant figures, the average value of density is 0.8196. The single best variable for correlation is $^3\chi_P$:

$$d = 0.7940 + 0.0169 \; ^3\chi_P, \quad r = 0.9247, \quad s = 0.0040, \quad N = 40$$

A five-variable function reduces the standard error by one-fourth:

$$d = 0.8251 - 0.0593/^1\chi - 0.0017 \; ^2\chi + 0.0144 \; ^3\chi_P$$

$$-0.0092 \; ^5\chi_P - 0.0058 \; ^3\chi_C^v,$$

$$r = 0.9633, \quad s = 0.0030, \quad N = 40$$

None of the residuals exceeds twice the standard error. Observed and calculated results are tabulated in Table XIII.

C. Aliphatic Ethers

There are only 13 ethers for which quality data are available. As with alcohols the best single variable is $^3\chi_P$:

$$d = 0.6963 + 0.4020 \; ^3\chi_P, \quad r = 0.8715, \quad s = 0.0082, \quad N = 13$$

The addition of $^5\chi_P$ decreases the error somewhat:

$$d = 0.6967 + 0.0373 \; ^3\chi_P + 0.0209 \; ^5\chi_P,$$

$$r = 0.9104, \quad s = 0.0072, \quad N = 13$$

Such a small data set does not warrant more than two variables. Observed and calculated values are given in Table XIV.

TABLE XIII

Observed and Calculated Values for Alcohol Density D_4^{20} Correlated with the Connectivity Function

Compound	D_4^{20}	
	Obs	Calc
1. 2-Propanol	0.78507	0.78642
2. 1-Propanol	0.80359	0.79962
3. 2-Methyl-1-propanol	0.80196	0.80530
4. 1-Butanol	0.80960	0.80841
5. 2-Methyl-2-butanol	0.80889	0.80707
6. 3-Methyl-2-butanol	0.81800	0.81763
7. 2-Pentanol	0.81030	0.81139
8. 3-Methyl-1-butanol	0.80918	0.81008
9. 2-Methyl-1-butanol	0.81930	0.81911
10. 1-Pentanol	0.81479	0.81333
11. 2,3-Dimethyl-2-butanol	0.82360	0.82390
12. 3,3-Dimethyl-2-butanol	0.81850	0.82390
13. 2-Methyl-2-pentanol	0.81341	0.80931
14. 2,2-Dimethyl-1-butanol	0.82834	0.82179
15. 3-Methyl-3-pentanol	0.82334	0.82485
16. 4-Methyl-2-pentanol	0.80750	0.81113
17. 2-Methyl-3-pentanol	0.82487	0.82510
18. 4-Methyl-1-pentanol	0.81310	0.81287
19. 3-Hexanol	0.82130	0.82148
20. 2-Ethyl-1-butanol	0.83345	0.82757
21. 1-Hexanol	0.81893	0.81835
22. 2,4-Dimethyl-2-pentanol	0.81030	0.81536
23. 2,4-Dimethyl-3-pentanol	0.82880	0.82864
24. 3-Ethyl-3-pentanol	0.83889	0.83801
25. 2,4,4-Trimethyl-2-pentanol	0.82250	0.81536
26. 2-Heptanol	0.81900	0.81837
27. 3-Heptanol	0.82270	0.82343
28. 4-Heptanol	0.81830	0.82314
29. 1-Heptanol	0.82242	0.82242
30. 2-Methyl-2-heptanol	0.81420	0.81592
31. 3-Methyl-3-heptanol	0.82820	0.82623
32. 4-Methyl-4-heptanol	0.82480	0.82494
33. 3-Ethyl-3-hexanol	0.83730	0.83693
34. 6-Methyl-1-heptanol	0.81760	0.82165
35. 2-Octanol	0.82160	0.82119
36. 2-Ethyl-1-hexanol	0.83280	0.83069
37. 1-Octanol	0.82700	0.82599
38. 2-Methyl-2-octanol	0.82100	0.81923
39. 4-Ethyl-4-heptanol	0.83500	0.83455
40. 3,7-Dimethyl-1-octanol	0.82850	0.82920

TABLE XIV

Observed and Calculated Values for Ether Density D_4^{20} Correlated with the Connectivity Function

		D_4^{20}	
Compound		Obs	Calc
1.	Diethyl ether	0.71378	0.72307
2.	Methyl-*tert*-butyl ether	0.74050	0.73626
3.	Ethyl-*n*-propyl ether	0.73860	0.73762
4.	*n*-Butylmethyl ether	0.74430	0.73762
5.	Ethyl-*tert*-butyl ether	0.74040	0.73400
6.	Methyl-*tert*-pentyl ether	0.77030	0.76809
7.	*sec*-Butylethyl ether	0.75030	0.75610
8.	*n*-Butylethyl ether	0.74900	0.74911
9.	Di-*n*-propyl ether	0.73600	0.74911
10.	Ethyl-*tert*-pentyl ether	0.76570	0.75553
11.	*n*-Butyl-isopropyl ether	0.75940	0.75741
12.	Ethyl-*n*-pentyl ether	0.76220	0.76105
13.	Di-*n*-butyl ester	0.76890	0.77298

D. Aliphatic Acids

In our fourth paper on molecular connectivity, the third class of compounds studied consists of aliphatic carboxylic acids [5]. Twenty were included. Correlation with reciprocal $^1\chi$ and $^3\chi_P$ yields good results:

$$d = 0.7546 + 0.4358/{^1\chi} + 0.0252\ {^3\chi_P}$$

$$r = 0.9831, \qquad s = 0.0137, \qquad N = 20$$

Observed and calculated results are given in Table XV.

V. Water Solubility of Organic Liquids

A physicochemical property that is of considerable practical importance is solubility. We consider here systems in which the solvent is water and the solutes are organic liquids.

Water solubility is a complex phenomenon. Intermolecular forces, statistical effects, and dissociation play important roles. For the present we shall exclude compounds that undergo significant ionization and give our attention to a variety of heteroatomic organic molecules.

TABLE XV

Observed and Calculated Values for Density of Aliphatic Acids Correlated with the Connectivity Function

Compound	D_4^{20}	
	Obs	Calc
1. Formic	1.2203	1.19654
2. Acetic	1.0493	1.07595
3. Propionic	0.9939	0.99751
4. 2-Methylpropionic	0.9479	0.96919
5. Butyric	0.9579	0.95219
6. 2-Methylbutanoic	0.9380	0.94712
7. 3-Methylbutanoic	0.9262	0.93101
8. 3,3-Dimethylbutanoic	0.9124	0.91795
9. Pentanoic	0.9392	0.92829
10. 2,2-Dimethylbutanoic	0.9276	0.94374
11. 2,3-Dimethylbutanoic	0.9275	0.93576
12. 2-Ethylbutanoic	0.9239	0.92682
13. 2-Methylpentanoic	0.9230	0.92526
14. 3-Methylpentanoic	0.9262	0.92147
15. 4-Methylpentanoic	0.9225	0.91679
16. Hexanoic	0.9357	0.91270
17. 2-Methylhexanoic	0.9194	0.91460
18. 4-Ethyl-4-methylbutanoic	0.9149	0.91345
19. Heptanoic	0.9198	0.90717
20. Octanoic	0.9088	0.90087

A principal driving force of any mixing process is the entropy gain achieved in the mixed state. If there were no heat effects (ΔH of mixing) associated with the solution process, then the solvent and solute would be completely miscible. Any decrease in enthalpy, evolution of heat, serves only to increase the driving force toward miscibility.

The degree of solubility then is determined by the solute–solvent interactions. When two liquids A and B are mixed, if the A–B interactions are similar in magnitude to the A–A and B–B interactions, there is little energy change, $\Delta H \approx 0$, and the solubility is high. If, however, the A–B interactions are significantly less than either the A–A or B–B intermolecular forces, complete mixing would simply increase the energy and offset the entropy gain. Hence, the solubility is kept low.

Low molecular weight alcohols are characterized by dipolar and hydrogen bonding forces not highly dissimilar to those of water. Hence, methanol through the propanols are completely miscible with water. However, as the hydrocarbon portion of the molecule increases, the solute–solvent interactions lessen in strength relative to solute–solute forces, as the weaker dispersion forces become more important. As a result, solubility decreases.

Although it is a complex phenomenon, solubility is characterized by the strength of intermolecular forces. To the extent that molecular connectivity supplies a model for such effects as dispersion and polarization, the connectivity method may be applicable to solubility studies. Polyfunctional compounds capable of hydrogen bonding by more than one mode may require some additional treatment.

In the present case we shall consider hydrocarbons, alcohols, ethers, and esters. There is a severe problem here concerning the quality of data. Little has been done to provide a rigorously scrutinized set of data as has been done for other equilibrium properties such as thermodynamic functions. Hence, somewhat arbitrary selections have been made

TABLE XVI

Observed and Predicted Values for Solubility of Alkanes Correlated with Connectivity

	Solubility (molal)	
Compound	Obs	Calc
1. n-Butane	2.43×10^{-3}	1.74×10^{-3}
2. Isobutane	2.83×10^{-3}	2.75×10^{-3}
3. n-Pentane	5.37×10^{-4}	4.91×10^{-4}
4. 2-Methylbutane	6.61×10^{-4}	7.06×10^{-4}
5. 3-Methylbutane	1.48×10^{-4}	1.82×10^{-4}
6. 2,2-Dimethylpropane	7.48×10^{-4}	1.40×10^{-3}
7. 2,2-Dimethylbutane	2.14×10^{-4}	3.38×10^{-4}
8. 2,4-Dimethylpentane	4.07×10^{-5}	8.13×10^{-5}
9. 2,2,4-Trimethylpentane	7.48×10^{-5}	3.91×10^{-5}
10. 2,2,5-Trimethylhexane	8.95×10^{-6}	1.10×10^{-5}
11. Cyclohexane	6.61×10^{-4}	3.95×10^{-4}
12. Methylcyclohexane	1.41×10^{-4}	1.46×10^{-4}
13. 1,2-Dimethylcyclohexane	5.38×10^{-5}	5.17×10^{-5}
14. Cycloheptane	3.05×10^{-4}	1.12×10^{-4}
15. Cyclooctane	7.05×10^{-5}	3.17×10^{-5}
16. n-Hexane	1.11×10^{-4}	1.39×10^{-4}
17. n-Heptane	2.93×10^{-5}	3.95×10^{-5}
18. n-Octane	5.79×10^{-6}	1.11×10^{-5}

in some cases and, in others, multiple listings of data have been used. Although such a method is less than ideal, it appears to be the only course at present [7]. The conclusions one might wish to draw from such studies are, of course, open to some reasonable questions.

A. Hydrocarbons

Two classes of hydrocarbons have been studied, alkanes and alkyl-substituted benzenes.

1. Alkanes

In our second paper on connectivity [3] we correlated alkane solubility with $^1\chi$ as follows:

$$\ln S = -1.505 - 2.533\,^1\chi, \qquad r = 0.958, \qquad s = 0.511, \qquad N = 18$$

Observed and predicted solubilities are given in Table XVI.

2. Alkyl-Substituted Benzenes

Data are available for only 13 alkyl-substituted benzenes. A variety of structure types are represented in this study. As might be expected,

TABLE XVII

Observed and Predicted Values of Logarithm of Solubility of Alkylbenzenes Correlated with Connectivity

		ln S	
Compound		Obs	Calc
1.	Benzene	1.64	1.79
2.	Toluene	2.25	2.26
3.	Ethylbenzene	2.84	2.84
4.	*o*-Xylene	2.78	2.71
5.	*m*-Xylene	2.76	2.72
6.	*p*-Xylene	2.73	2.72
7.	*n*-Propylbenzene	3.34	3.38
8.	Isopropylbenzene	3.38	3.24
9.	1,2,4-Trimethylbenzene	3.32	3.18
10.	*n*-Butylbenzene	3.94	3.92
11.	*sec*-Butylbenzene	3.67	3.89
12.	*tert*-Butylbenzene	3.60	3.56
13.	*tert*-Amylbenzene	4.15	4.18

correlation with $^1\chi$ alone is not good but the addition of $^1\chi^v$ provides a satisfactory relationship:

$$\ln S = -1.045\ ^1\chi + 2.129\ ^1\chi^v + 0.671$$

$$r = 0.9890, \qquad s = 0.111, \qquad N = 13$$

Observed and predicted values are given in Table XVII.

B. Aliphatic Alcohols

Somewhat more data are available for aliphatic alcohols than for alkanes. A set of 51 alcohols was studied with $^1\chi$ [3]:

$$\ln S = 6.702 - 2.666\ ^1\chi, \qquad r = 0.978, \qquad s = 0.455, \qquad N = 51 \quad (3)$$

This correlation can be improved significantly by the introduction of the empirical parameter c_{OH} as discussed in conjunction with alcohol boiling point. At the same time the set of data for alkanes may be merged with the alcohols by introducing a dummy variable Q, which has the value 1 for an alcohol and 0 for a hydrocarbon. The result is

$$\ln S = -1.273 - 2.611\ ^1\chi - 4.405c_{OH} + 10.42Q \qquad (4)$$

$$r = 0.994, \qquad s = 0.357, \qquad N = 68$$

For the alcohols alone, the equation is

$$\ln S = 9.204 - 2.630\ ^1\chi - 4.390c_{OH}$$

$$r = 0.9914, \qquad s = 0.289, \qquad N = 50$$

The results for this equation are tabulated in Table XVIII.

This equation can be used to predict the solubility of higher molecular weight alcohols as shown in Table XIX. The high quality of these results suggests the practical potential use of this method.

In order to make a meaningful application of the connectivity function to alcohol solubility, we purged from the list the least reliable data points. The remaining data serve as a good test of the method. The addition of $^1\chi^v$ to $^1\chi$ decreases the standard error for this data set from 0.443 to 0.234:

$$\ln S = 9.417 - 11.266\ ^1\chi + 8.643\ ^1\chi^v$$

$$r = 0.9945, \qquad s = 0.234, \qquad N = 38$$

Additional extended terms continue to decrease the standard error. A

TABLE XVIII

*Observed and Predicted Logarithm of Solubility of Aliphatic Alcohols
Correlated with Connectivity*

		ln S	
Compound		Obs	Calc
1.	*n*-Butanol	0.00598	−0.2480
2.	2-Methylpropanol	0.02273	0.1332
3.	2-Butanol	0.06578	0.7040
4.	*n*-Pentanol	−1.347	−1.562
5.	3-Methylbutanol	−1.167	−1.181
6.	2-Methylbutanol	−1.058	−1.281
7.	2-Pentanol	−0.6348	−0.6107
8.	3-Pentanol	−0.4861	−0.7107
9.	3-Methyl-2-butanol	−0.4049	−0.2741
10.	2-Methyl-2-butanol	0.3386	0.2742
11.	*n*-Hexanol	−2.790	−2.877
12.	2-Hexanol	−1.995	−1.925
13.	3-Hexanol	−1.832	−2.025
14.	3-Methyl-3-pentanol	−0.8301	−1.200
15.	2-Methyl-2-pentanol	−1.117	−1.040
16.	2-Methyl-3-pentanol	−1.609	−1.688
17.	3-Methyl-2-pentanol	−1.639	−1.688
18.	2,3-Dimethyl-2-butanol	−0.8509	−0.7302
19.	3,3-Dimethylbutanol	−2.590	−1.949
20.	3,3-Dimethyl-2-butanol	−1.410	−1.068
21.	4-Methylpentanol	−2.282	−2.496
22.	4-Methyl-2-pentanol	−1.814	−1.544
23.	2-Ethylbutanol	−2.787	−2.696
24.	Cyclohexanol	−0.9597	−0.9368
25.	*n*-Heptanol	−4.166	−4.192
26.	2-Methyl-2-hexanol	−2.473	−2.355
27.	3-Methyl-3-hexanol	−2.263	−2.515
28.	3-Ethyl-3-pentanol	−1.917	−2.676
29.	2,3-Dimethyl-2-pentanol	−2.002	−2.145
30.	2,3-Dimethyl-3-pentanol	−1.937	−2.205
31.	2,4-Dimethyl-2-pentanol	−2.145	−1.974
32.	2,4-Dimethyl-3-pentanol	−2.801	−2.669
33.	2,2-Dimethyl-3-pentanol	−2.643	−2.483
34.	3-Heptanol	−3.194	−3.340
35.	4-Heptanol	−3.196	−3.340
36.	*n*-Octanol	−5.401	−5.507
37.	2,2,3-Trimethyl-3-pentanol	−2.931	−3.012
38.	2-Octanol	−4.755	−4.555
39.	2-Ethylhexanol	−4.996	−5.325

TABLE XVIII—Continued

Compound	ln S	
	Obs	Calc
40. *n*-Nonanol	−6.907	−6.822
41. 2-Nonanol	−6.319	−5.869
42. 3-Nonanol	−6.119	−5.969
43. 4-Nonanol	−5.952	−5.969
44. 5-Nonanol	−5.744	−5.969
45. 2,6-Dimethyl-3-heptanol	−5.776	−5.207
46. 3,5-Dimethyl-4-heptanol	−5.298	−5.499
47. 1,1-Diethylpentanol	−5.572	−5.305
48. 7-Methyloctanol	−5.744	−6.440
49. 3,5,5-Trimethylhexanol	−5.769	−5.612
50. *n*-Decanol	−8.517	−8.136

four-variable function brings the standard error under 0.2:

$$\ln S = 8.996 - 11.627 \, ^1\chi + 9.507 \, ^1\chi^v - 0.300 \, ^2\chi - 2.385 \, ^4\chi_C^v$$

$$r = 0.9966, \qquad s = 0.189, \qquad N = 38$$

Observed and predicted values are shown in Table XX.

C. Aliphatic Ethers

Reliable data are available for only a few ethers. As Amidon *et al.* have recently reported [7], more than one value appears in the literature. Such multiple listings are retained in the correlation analysis.

Correlation with $^1\chi$ is quite good:

$$\ln S = 6.027 - 2.772 \, ^1\chi, \qquad r = 0.9895, \quad s = 0.242, \quad N = 22$$

Observed and predicted values are listed in Table XXI.

In an additional study the above alcohols and ethers were combined for a total of 61 observations. Correlation with $^1\chi$ gave $r = 0.9678$ and $s = 0.556$. However, addition of $^1\chi^v$ brought significant improvement:

$$\ln S = 8.834 - 9.247 \, ^1\chi + 6.583 \, ^1\chi^v$$

$$r = 0.9916, \qquad s = 0.285, \qquad N = 61$$

Such an excellent correlation is quite interesting in view of the difference in the intermolecular forces between these compounds. In fact, the

TABLE XIX

Observed and Predicted Solubilities for Three Solid Alcohols Based on Correlation with Connectivity

| Compound | S | | |
	Obs[a]	Calc[b]	Calc[c]
n-Tetradecanol	2.84×10^{-6}	2.12×10^{-6}	2.43×10^{-6}
n-Pentadecanol	1.02×10^{-6}	0.56×10^{-6}	0.69×10^{-6}
n-Hexadecanol	4.55×10^{-7}	1.47×10^{-7}	1.94×10^{-7}

[a] Taken from G. L. Amidon, S. H. Yalkowski, and S. Leung, *J. Pharm. Sci.* **63,** 1858 (1974).
[b] Based on Eq. (3).   [c] Based on Eq. (4).

use of a dummy variable δ^v to describe hydrogen bonding, as was done with boiling points, brought only minor improvement: $r = 0.9913$, $s = 0.274$. This result serves to underscore further the range of applicability of the connectivity method.

D.   Aliphatic Esters

For aliphatic esters 38 observations have been used for 25 compounds. Correlation with $^1\chi$ is quite good: $r = 0.9833$, $s = 0.354$. The addition of $^1\chi^v$ brings about a modest improvement:

$$\ln S = 9.128 - 5.494 \, ^1\chi + 2.965 \, ^1\chi^v$$

$$r = 0.9853, \qquad s = 0.337, \qquad N = 38$$

Observed [7] and predicted values are given in Table XXII.

Although additional extended terms can provide some further improvement, we feel that the quality of the data do not warrant such a study at this time.

E.   A Combined Study of Alcohols, Ethers, and Esters

The combined set of alcohols, ethers, and esters provides a group of 98 compounds of carbon, hydrogen, and oxygen that have several differences in structure and intermolecular forces. Only the alcohols possess the OH groups that participate in hydrogen bonding. Only the

TABLE XX

Observed and Predicted Logarithm of Solubility of Aliphatic Alcohols Correlated with Connectivity

		ln S	
Compound		Obs	Calc
1.	2-Methyl-1-propanol	0.02273	−0.0733
2.	2-Butanol	0.06578	0.6093
3.	1-Butanol	0.00598	−0.2444
4.	2-Methyl-2-butanol	0.3386	0.0655
5.	3-Methyl-2-butanol	−0.4050	−0.3876
6.	2-Pentanol	−0.6349	−0.5648
7.	3-Methyl-1-butanol	−1.1680	−1.247
8.	2-Methyl-1-butanol	−1.058	−1.250
9.	1-Pentanol	−1.347	−1.411
10.	2,3-Dimethyl-2-butanol	−0.8510	−0.928
11.	2-Methyl-2-pentanol	−1.118	−1.113
12.	3,3-Dimethyl-1-butanol	−2.590	−2.202
13.	3-Methyl-2-pentanol	−0.8301	−1.110
14.	4-Methyl-2-pentanol	−1.814	−1.571
15.	3-Methyl-2-pentanol	−1.640	−1.571
16.	2-Methyl-3-pentanol	−1.605	−1.571
17.	4-Methyl-1-pentanol	−2.283	−2.413
18.	2-Hexanol	−1.995	−1.731
19.	3-Hexanol	−1.833	−1.741
20.	1-Hexanol	−2.790	−2.765
21.	2,2-Dimethyl-3-pentanol	−2.644	−2.521
22.	2,3-Dimethyl-2-pentanol	−2.002	−2.115
23.	2,3-Dimethyl-3-pentanol	−1.938	−2.109
24.	2-Methyl-2-hexanol	−2.474	−2.280
25.	3-Methyl-3-hexanol	−2.264	−2.289
26.	3-Heptanol	−3.194	−3.252
27.	4-Heptanol	−3.197	−3.260
28.	2,2,3-Trimethyl-3-pentanol	−2.932	−3.056
29.	1-Heptanol	−4.167	−4.174
30.	2-Octanol	−4.756	−4.659
31.	1-Octanol	−5.402	−5.551
32.	3,5,5-Trimethyl-1-hexanol	−5.770	−5.888
33.	3,5-Dimethyl-4-heptanol	−5.298	−5.171
34.	3-Nonanol	−6.119	−6.007
35.	4-Nonanol	−5.952	−6.060
36.	5-Nonanol	−5.745	−5.906
37.	1-Nonanol	−6.908	−6.928
38.	1-Decanol	−8.517	−8.305

TABLE XXI

Observed and Predicted Values for Logarithm of Solubility of Ethers
Correlated with the Connectivity Function

		ln S	
Compound		Obs	Calc
1.	Dimethyl ether	1.7715	1.8483
2.	Isopropyl methyl ether	−0.1381	−0.2053
3.	Isopropyl methyl ether	−0.0650	−0.2053
4.	Diethyl ether	−0.5499	−0.4434
5.	Diethyl ether	−0.2536	−0.4434
6.	Diethyl ether	−0.5499	−0.4434
7.	Methyl propyl ether	−0.6198	−0.7883
8.	Methyl propyl ether	−0.8770	−0.7883
9.	Methyl propyl ether	−0.6198	−0.7883
10.	Ethyl isopropyl ether	−1.2909	−1.3514
11.	Ethyl isopropyl ether	−1.2909	−1.3514
12.	Methyl isobutyl ether	−2.0714	−1.8493
13.	Methyl isobutyl ether	−2.0714	−1.8493
14.	Methyl *sec*-butyl ether	−1.7037	−1.8097
15.	Methyl *sec*-butyl ether	−1.7037	−1.8097
16.	Butyl methyl ether	−2.3025	−2.2793
17.	Butyl methyl ether	−2.3025	−2.2793
18.	Isopropyl propyl ether	−3.0856	−2.8424
19.	Isopropyl propyl ether	−3.0856	−2.8424
20.	Dipropyl ether	−3.3639	−3.4254
21.	Dipropyl ether	−3.3639	−3.4254
22.	Dibutyl ether	−6.2606	−6.4074

esters possess the highly polar carbonyl function. Yet a simple connectivity function provides excellent correlation:

$$\ln S = 9.240 - 5.423\ {}^1\chi + 2.819\ {}^1\chi^v - 0.382d$$

$$r = 0.9843, \qquad s = 0.357, \qquad N = 98$$

The variable d, which may provide some discrimination for the alcohol hydrogen bonding, is equal to δ_0^v for alcohols and ethers, but $d = 0$ for esters.

This simple but effective function may be useful in further solubility studies, which might include ketones, aldehydes, halides, and aromatic compounds.

TABLE XXII

Observed and Calculated Values for Logarithm of Solubility of Esters Correlated with the Connectivity Function

		ln S	
Compound		Obs	Calc
1.	Methyl formate	1.0152	1.219
2.	Ethyl formate	0.1739	0.214
3.	Ethyl formate	−0.3453	0.214
4.	Propyl formate	−1.1332	−1.050
5.	Propyl formate	−1.1744	−1.050
6.	1-Butyl formate	−2.3025	−1.950
7.	Butyl formate	−2.7333	−2.315
8.	1-Pentyl formate	−3.5000	−3.214
9.	Methyl acetate	1.1908	0.560
10.	Methyl acetate	0.9242	0.560
11.	Ethyl acetate	−0.0921	−0.445
12.	Ethyl acetate	−0.0693	−0.445
13.	Isopropyl acetate	−1.1940	−1.230
14.	Isopropyl acetate	−1.2447	−1.230
15.	Propyl acetate	−1.7037	−1.710
16.	Propyl acetate	−1.7259	−1.710
17.	Isobutyl acetate	−2.8490	−2.609
18.	Butyl acetate	−3.1535	−2.974
19.	Isopentyl acetate	−4.3981	−3.874
20.	Pentyl acetate	−4.2830	−4.239
21.	Hexyl acetate	−4.7205	−5.503
22.	Methyl propionate	−0.3453	−0.734
23.	Methyl propionate	−0.3900	−0.734
24.	Ethyl propionate	−1.4740	−1.739
25.	Ethyl propionate	−1.6660	−1.739
26.	Isopropyl propionate	−2.9700	−2.524
27.	Propyl propionate	−3.0856	−3.003
28.	Propyl propionate	−2.9917	−3.003
29.	Butyl propionate	−4.3050	−4.268
30.	Isopentyl propionate	−5.0880	−5.168
31.	Propyl propionate	−5.1814	−5.532
32.	Methyl butyrate	−1.9449	−1.998
33.	Methyl butyrate	−1.9877	−1.998
34.	Ethyl butyrate	−2.9355	−3.003
35.	Isopropyl butyrate	−4.4654	−3.788
36.	Propyl butyrate	−4.4228	−4.268
37.	Propyl butyrate	−4.3900	−4.268
38.	Ethyl heptanoate	−6.3034	−6.800

VI. Partition Coefficient

Closely related to solubility is the partitioning of a solute between two immiscible liquid phases. Rigorously defined, the partition coefficient P is the ratio of the activities of the solute in the two phases. In practice, however, a ratio of molar concentrations is often used.

We confine our discussions here to the water/octanol-1 system because of the greater availability of data. Furthermore, this partition coefficient is currently used as a model for certain drug properties [8]. The partition coefficient is defined as

$$P = c_O/c_W$$

where c_O is the concentration in the octanol phase and c_W the concentration in the water phase.

The principal physical effects in partitioning relate to the nature and magnitude of the intermolecular forces between the solute and the two solvents. Entropy effects attributed to mixing in each phase are approximately equal except in cases in which the degree of association of solute molecules varies between the two solvents. The partition coefficient P is large for substances whose intermolecular forces are similar in magnitude and nature to those for octanol: $\log P = 3.15$ for nonanol. For substances with intermolecular forces more similar to water, P becomes quite small, even significantly less than zero: $\log P = -0.66$ for methanol and -0.57 for methylamine.

TABLE XXIII

Correlation of $^1\chi^v$ with Partition Coefficient P: Regression Coefficients and Statistical Data

Compound class	Slope	Intercept	r	s	N
Alcohols	0.966	−1.53	0.997	0.151	49
Ethers	0.964	−1.30	0.976	0.080	12
Ketones	0.982	−1.16	0.993	0.094	16
Acids	0.927	−1.14	0.996	0.122	9
Esters	0.996	−1.71	0.999	0.060	24
Amines	0.977	−1.51	0.979	0.179	28
Hydrocarbons	0.884	0.406	0.975	0.160	45
All classes except hydrocarbons	0.950	−1.48	0.986	0.152	138

For substances dissimilar to water and octanol there are also considerable variations in the magnitude of P. For methyl acetate, log $P = 0.23$, whereas for butyl pentanoate, log $P = 3.23$. The possible variations in the type and size of the intermolecular forces between solutes and the two solvents present a most complex picture. Although there are typical homologous series relationships, effects due to skeletal branching and functional group position do not reflect the same pattern as with heat of formation or boiling point. Isopentanol, 2-methylbutanol, 1-methylbutanol, and 3-pentanol all possess the same value for log P: 1.14.

These observations suggest a complex relation between partition coefficient and the topology of the molecule. Factors not strongly dependent on molecular topology may play a significant role, especially in comparing chemically different molecules such as aliphatic alcohols and substituted aromatic amines. Nonetheless, some significant correlations have been obtained between the connectivity function and log P.

A. A Study of Mixed Compounds

In an initial study on partition coefficient [9], a mixed list of 138 compounds was studied. Included in the list are 49 alcohols, 12 ethers, 16 ketones, 9 carboxylic acids, 25 esters, and 27 amines. Correlation with $^1\chi$ is unsatisfactory, producing a set of nearly parallel lines relating to the degree of unsaturation in the molecules.

When the use valence deltas is introduced only for those atoms involved in multiple bonding,* the use of the single variable $^1\chi^v$ produces the following result:

$$\log P = 1.48 + 0.950 \, ^1\chi^v, \qquad r = 0.986, \quad s = 0.152, \quad N = 138 \quad (5)$$

Correlation with $^1\chi^v$ for each class of compounds yields a somewhat smaller standard error, as shown in Table XXIII. The slope for each equation is statistically indistinguishable from unity although the average value is slightly less than 1.00. The variation in intercept, -1.16 to -1.71, is the major cause in the decrease in r when all classes are regressed together. Even so, the correlation for the whole set is quite satisfactory. Observed and calculated values for Eq. (5) are given in Table XXIV.

Hydrocarbons are not included in this list because the slope and intercept are significantly different:

$$\log P = 0.406 + 0.884 \, ^1\chi^v, \qquad r = 0.975, \quad s = 0.160, \quad N = 45$$

The correlation is quite good for a list that includes cyclics and

* The relationship $\delta^v = Z^v - h$ is not used for heteroatoms.

TABLE XXIV

Observed and Predicted Values for Partition Coefficient of Several Classes of Compounds Correlated with the Connectivity Function

		log P		
Compound		Obs[a]	Calc[b]	Calc[c]
1.	Methanol	−0.66	−0.57	−0.68
2.	Ethanol	−0.32	−0.17	−0.11
3.	*n*-Propanol	0.34	0.31	0.38
4.	*n*-Butanol	0.88	0.80	0.87
5.	*n*-Pentanol	1.40	1.28	1.34
6.	*n*-Hexanol	1.84	1.76	1.82
7.	*n*-Heptanol	2.34	2.25	2.29
8.	*n*-Octanol	2.84	2.73	2.77
9.	*n*-Nonanol	3.15	3.21	3.25
10.	Isopropanol	0.14	0.14	0.16
11.	Isobutanol	0.61	0.66	0.61
12.	*tert*-Butanol	0.37	0.40	0.51
13.	Isopentanol	1.14	0.94	1.09
14.	2-Methylbutanol	1.14	1.17	1.15
15.	1-Methylbutanol	1.14	1.14	1.15
16.	3-Pentanol	1.14	1.17	1.28
17.	3-Methyl-2-butanol	0.91	1.14	0.94
18.	2-Methyl-2-butanol	0.89	0.94	1.01
19.	2,2-Dimethyl-1-propanol	1.36	0.94	1.08
20.	2-Hexanol	1.61	1.62	1.64
21.	3-Hexanol	1.61	1.66	1.68
22.	3-Methyl-3-pentanol	1.39	1.48	1.50
23.	2-Methyl-2-pentanol	1.39	1.42	1.43
24.	2-Methyl-3-pentanol	1.41	1.54	1.42
25.	3-Methyl-2-pentanol	1.41	1.54	1.45
26.	4-Methyl-2-pentanol	1.41	1.48	1.35
27.	2,3-Dimethyl-2-butanol	1.17	1.31	1.23
28.	3,3-Dimethyl-1-butanol	1.86	1.42	1.56
29.	3,3-Dimethyl-2-butanol	1.19	1.31	1.35
30.	2-Methyl-2-hexanol	1.87	1.90	1.94
31.	3-Methyl-3-hexanol	1.87	1.96	1.95
32.	3-Ethyl-3-pentanol	1.87	2.02	1.97
33.	2,3-Dimethyl-2-pentanol	1.67	1.83	1.72
34.	2,3-Dimethyl-3-pentanol	1.67	1.85	1.71
35.	2,4-Dimethyl-2-pentanol	1.67	1.77	1.61
36.	2,4-Dimethyl-3-pentanol	1.71	1.74	1.63
37.	2,2-Dimethyl-3-pentanol	1.69	1.83	1.81
38.	2,2,3-Trimethyl-3-pentanol	1.99	2.15	2.05
39.	Cyclohexanol	1.23	1.26	1.53

TABLE XXIV—Continued

		log *P*	
Compound	Obs[a]	Calc[b]	Calc[c]
40. 4-Penten-1-ol	1.04	0.90	0.88
41. 3-Penten-2-ol	0.81	0.78	0.71
42. 1-Penten-3-ol	0.81	0.82	0.76
43. 1-Hexen-3-ol	1.31	1.30	1.15
44. 2-Hexen-4-ol	1.31	1.30	1.26
45. 2-Methyl-4-penten-3-ol	1.11	1.18	1.00
46. Benzyl alcohol	1.10	0.85	0.92
47. 2-Phenylethanol	1.36	1.34	1.40
48. 3-Phenyl-1-propanol	1.88	1.82	1.88
49. Diphenylcarbinol	2.67	2.30	2.56
50. Diethylether	1.03	1.02	1.02
51. Methyl butyl ether	1.53	1.50	1.50
52. Methyl *sec*-butyl ether	1.33	1.36	1.37
53. Methyl isobutyl ether	1.33	1.40	1.30
54. Methyl *tert*-butyl ether	1.06	1.16	1.01
55. Ethyl propyl ether	1.53	1.50	1.50
56. Ethyl isopropyl ether	1.33	1.36	1.30
57. Di-*n*-propyl ether	2.03	1.99	1.99
58. *n*-Propyl isopropyl ether	1.83	1.85	1.79
59. Methyl *n*-propyl ether	1.03	1.02	1.02
60. Methyl isopropyl ether	0.83	0.88	0.82
61. Ethyl cyclopropyl ether	1.24	1.04	1.47
62. Ethyl formate	0.23	0.30	0.22
63. *n*-Propyl formate	0.73	0.80	0.73
64. Methyl acetate	0.23	0.19	0.35
65. Ethyl acetate	0.73	0.69	0.73
66. *n*-Propyl acetate	1.23	1.19	1.16
67. Isopropyl acetate	1.03	1.05	0.96
68. *n*-Butyl acetate	1.73	1.69	1.80
69. *sec*-Butyl acetate	1.53	1.54	1.50
70. Methyl propionate	0.73	0.75	0.73
71. Methyl butyrate	1.23	1.25	1.17
72. Ethyl hexanoate	2.73	2.74	2.68
73. Ethyl heptanoate	3.23	3.24	3.19
74. Ethyl octanoate	3.73	3.74	3.70
75. Ethyl nonanoate	4.23	4.24	4.22
76. Ethyl decanoate	4.73	4.74	4.73
77. Ethyl propionate	1.23	1.25	1.25
78. Ethyl butyrate	1.73	1.75	1.63
79. Ethyl isobutyrate	1.53	1.63	1.63
80. Pentyl acetate	2.23	2.19	2.31
81. Butyl pentanoate	3.23	3.24	3.22

TABLE XXIV—Continued

		log P	
Compound	Obs[a]	Calc[b]	Calc[c]
82. Benzyl acetate	1.96	1.74	2.00
83. Methyl 3-phenylpropionate	2.32	2.30	2.34
84. Methyl 2-phenylacetate	1.83	1.80	1.80
85. Methyl 4-phenylbutyrate	2.77	2.80	2.88
86. Acetic acid	−0.17	−0.15	−0.31
87. Propionic acid	0.25	0.37	0.30
88. Butyric acid	0.79	0.83	0.83
89. Hexanoic acid	1.88	1.76	1.89
90. Decanoic acid	4.09	4.07	4.02
91. 2-Phenylacetic acid	1.41	1.34	1.43
92. 3-Phenylpropionic acid	1.84	1.81	1.96
93. 4-Phenylbutyric acid	2.42	2.27	2.50
94. 2,2-Diphenylacetic acid	3.05	3.26	2.95
95. Methylamine	−0.57	−0.54	−0.65
96. Ethylamine	−0.13	−0.13	−0.16
97. *n*-Propylamine	0.48	0.36	0.33
98. *n*-Butylamine	0.75	0.84	0.82
99. *n*-Pentylamine	1.49	1.33	1.31
100. *n*-Hexylamine	1.98	1.82	1.81
101. *n*-Heptylamine	2.57	2.31	2.30
102. Isobutylamine	0.73	0.70	0.71
103. *sec*-Butylamine	0.74	0.70	0.71
104. 2-Aminooctane	2.82	2.69	2.74
105. Cyclohexylamine	1.49	1.31	1.28
106. Isopropylamine	0.26	0.18	0.74
107. Methylethylamine	0.15	0.36	0.23
108. Di-*n*-propylamine	1.67	1.83	1.91
109. Triethylamine	1.44	1.76	1.28
110. Di-*n*-butylamine	2.68	2.80	2.69
111. Diethylamine	0.57	0.84	0.72
112. *n*-Propyl-*n*-butylamine	2.12	2.31	2.20
113. Methyl-*n*-butylamine	1.33	1.33	1.22
114. Piperidine	0.85	0.93	0.81
115. Ethyl-isopropylamine	0.93	1.19	1.17
116. *n*-Propyl-*sec*-butylamine	1.91	2.20	2.04
117. *n*-Propyl-isobutylamine	2.07	1.77	2.15
118. Trimethylamine	0.27	0.18	0.15
119. Dimethyl-*n*-butylamine	1.70	1.68	1.46
120. Dimethylbenzylamine	1.98	1.74	1.94
121. Benzylamine	1.09	0.90	1.22
122. Phenethylamine	1.41	1.39	1.65
123. Acetone	−0.21	−0.28	−0.18

TABLE XXIV—Continued

		log P	
Compound	Obs[a]	Calc[b]	Calc[c]
124. 2-Butanone	0.29	0.27	0.31
125. 2-Pentanone	0.79	0.76	0.77
126. 3-Pentanone	0.79	0.82	0.81
127. 3-Methyl-2-butanone	0.59	0.64	0.59
128. 2-Hexanone	1.29	1.25	1.23
129. 3-Hexanone	1.29	1.31	1.28
130. 3-Methyl-2-pentanone	1.09	1.17	1.07
131. 4-Methyl-2-pentanone	1.09	1.11	1.07
132. 2-Methyl-3-pentanone	1.09	1.20	1.11
133. 2-Heptanone	1.79	1.74	1.78
134. 3-Heptanone	1.79	1.80	1.81
135. 2,4-Dimethyl-3-pentanone	1.39	1.57	1.41
136. 5-Nonanone	2.79	2.78	2.80
137. 2-Nonanone	2.79	2.72	2.82
138. Acetophenone	1.58	1.35	1.54

[a] Partition coefficients in an *n*-octanol/water system were taken from the following sources: C. Hansch, J. E. Quinlan, and G. L. Lawrence, *J. Org. Chem.* **33**, 347(1968); G. G. Nys and R. F. Rekker, *Chem. Therapeut.* **5**, 521(1973); A. Leo, C. Hansch, and D. Elkins, *Chem. Rev.* **71**, 525(1971).

[b] Calculated from Eq. (3) for the whole set of data, 138 compounds.

[c] Calculated from Eqs. (4)–(9) for each class individually, as discussed in the text.

noncyclics, alkanes, alkenes, alkynes, substituted benzenes, naphthalene, and phenanthrene. Observed and calculated results are shown in Table XXV.

We will apply the connectivity function to these compounds by considering each class individually.

B. Aliphatic Alcohols

For the 49 alcohols in the list, $^1\chi^v$ gives significant correlation:

$$\log P = -0.985 + 0.860\ ^1\chi^v, \qquad r = 0.9612, \quad s = 0.195, \quad N = 42$$

The use of $^1\chi^v$ together with $^1\chi$ brings a modest improvement: $r = 0.9655$, $s = 0.186$. A five-variable connectivity function decreases the

TABLE XXV

*Observed and Predicted Values for Partition Coefficient of Hydrocarbons
Correlated with Connectivity*

		log P	
Compound		Obs	Calc
1.	*n*-Pentane	2.50	2.54
2.	2-Methylbutane	2.30	2.41
3.	2-Methylpentane	2.80	2.85
4.	3-Methylpentane	2.80	2.88
5.	*n*-Hexane	3.00	2.98
6.	*n*-Heptane	3.50	3.42
7.	2,4-Dimethylpentane	3.10	3.17
8.	*n*-Octane	4.00	3.89
9.	Cyclopentane	2.05	2.17
10.	Cyclohexane	2.46	2.62
11.	Methylcyclopentane	2.35	2.52
12.	Cycloheptane	2.87	3.06
13.	Methylcyclohexane	2.76	2.96
14.	Cyclooctane	3.28	3.50
15.	1,2-Dimethylcyclohexane	3.06	3.33
16.	1-Pentyne	1.98	2.04
17.	1-Hexyne	2.48	2.48
18.	1-Heptyne	2.98	2.93
19.	1-Octyne	3.48	3.37
20.	1-Nonyne	3.98	3.81
21.	1,8-Nonadiyne	3.46	3.31
22.	1,6-Heptadiyne	2.46	2.42
23.	1-Pentene	2.20	2.20
24.	2-Pentene	2.20	2.20
25.	1-Hexene	2.70	2.64
26.	2-Heptene	3.20	3.08
27.	1-Octene	3.70	3.52
28.	4-Methyl-1-pentene	2.50	2.51
29.	1,6-Heptadiene	2.90	2.73
30.	1,5-Hexadiene	2.40	2.29
31.	1,4-Pentadiene	1.90	1.85
32.	Cyclopentene	1.75	1.86
33.	Cyclohexene	2.16	2.31
34.	Cycloheptene	2.57	2.75
35.	Toluene	2.73	2.54
36.	Ethylbenzene	3.15	2.98
37.	Isopropylbenzene	3.66	3.23
38.	*n*-Propylbenzene	3.68	3.42
39.	Diphenylmethane	4.14	4.41
40.	1,2-Diphenylethane	4.79	4.85
41.	Biphenyl	4.04	4.01
42.	*p*-Xylene	3.25	2.90
43.	Benzene	2.13	2.17
44.	Naphthalene	3.37	3.42
45.	Phenanthrene	4.81	4.66

standard error by one-third:

$$\log P = -1.236 + 1.248 \, {}^1\chi^v - 0.2153 \, {}^2\chi^v - 0.2046 \, {}^4\chi^v_P$$
$$+ 1.079 \, {}^3\chi^v_C - 0.1399 \, {}^3\chi_P. \tag{6}$$
$$r = 0.9858, \qquad s = 0.124, \qquad N = 49$$

Observed and calculated values are given in Table XXIV.

C. Aliphatic Ethers

The variable ${}^1\chi$ is the best correlation variable for ethers:

$$\log P = -1.411 + 0.988 \, {}^1\chi, \qquad r = 0.9680, \quad s = 0.091, \quad N = 12$$

The addition of ${}^1\chi^v$ does not give significant improvement. The use of ${}^2\chi$ does decrease the standard error about 10%:

$$\log P = -1.346 + 1.038 \, {}^1\chi - 0.103 \, {}^2\chi \tag{7}$$
$$r = 0.9762, \qquad s = 0.083, \qquad N = 12$$

Because the ether linkage plays a constant structural role, it appears that no valence terms are required at this level of description. Observed and calculated values are given in Table XXIV.

D. Aliphatic Ketones

There are 16 ketones in the list and ${}^1\chi^v$ is the best correlation variable:

$$\log P = -1.468 + 0.985 \, {}^1\chi^v, \qquad r = 0.9929, \quad s = 0.098, \quad N = 16$$

The addition of ${}^1\chi$ improves the quality:

$$\log P = -1.624 + 0.692 \, {}^1\chi^v + 0.290 \, {}^1\chi$$
$$r = 0.9967, \qquad s = 0.068, \qquad N = 16$$

However, the use of ${}^1\chi^v$ and ${}^6\chi_P$ yields a correlation with $r = 0.9983$ and $s = 0.049$, whereas the use of ${}^6\chi^v_C$ decreases the standard error even more:

$$\log P = -1.270 + 0.909 \, {}^1\chi^v + 0.572 \, {}^6\chi_P - 0.150 \, {}^6\chi^v_C \tag{8}$$
$$r = 0.9993, \qquad s = 0.031, \qquad N = 16$$

Observed and predicted values are given in Table XXIV.

E.   Aliphatic Carboxylic Acids

There are only nine acids in the list and $^1\chi^v$ does not yield a satisfactory standard error:

$$\log P = -1.087 + 0.904\ ^1\chi^v, \qquad r = 0.9708, \quad s = 0.346, \quad N = 9$$

The addition of $^1\chi$ causes significant improvement in the relation:

$$\log P = -0.859 + 1.615\ ^1\chi^v - 0.550\ ^1\chi \tag{9}$$

$$r = 0.9979, \qquad s = 0.099, \qquad N = 9$$

Observed and predicted values are given in Table XXIV.

F.   Esters

For the 24 esters in the list, correlation with $^1\chi^v$ gives a large error:

$$\log P = -0.935 + 0.913\ ^1\chi^v, \qquad r = 0.9794, \quad s = 0.251, \quad N = 24$$

The addition of $^1\chi$ does not bring significant improvement. However, the addition of $^4\chi_{PC}$ cuts the standard error to 0.181. The further addition of $^0\chi^v$ brings about excellent correlation:

$$\log P = -1.778 + 0.256\ ^1\chi^v + 0.541\ ^0\chi^v - 0.535\ ^4\chi_{PC} \tag{10}$$

$$r = 0.9988, \qquad s = 0.064, \qquad N = 24$$

Observed and predicted results are listed in Table XXIV.

G.   Aliphatic Amines

There are 28 amines in the list, including primary and secondary amines with a few tertiary compounds. Correlation with $^1\chi$ or with $^1\chi^v$ and $^1\chi$ together is not satisfactory, giving a standard error of 0.31 for the combination of variables.

However, the use of other connectivity variables together with the valence delta δ_N^v gives a good correlation:

$$\log P = -1.001 + 0.696\ ^0\chi^v - 0.566\ ^6\chi_C^v - 0.248\delta_N^v$$

$$r = 0.9797, \qquad s = 0.184, \qquad N = 28$$

Observed and calculated values are given in Table XXIV.

H.   Alcohols and Ethers

The alcohols and ethers may be taken together for a study of mixed classes of compounds. Correlation with $^1\chi^v$ yields

$$\log P = -0.839 + 0.828\ ^1\chi^v, \qquad r = 0.9415, \quad s = 0.219, \quad N = 61$$

When δ_0^v and $^3\chi_C$ are added, the fit is improved:

$$\log P = -2.529 + 1.042 \, ^1\chi^v - 0.219 \, ^3\chi_C + 0.274\delta_0^v$$

$$r = 0.9725, \qquad s = 0.154, \qquad N = 61$$

The addition of $^4\chi_P^v$ and $^2\chi$ continues to improve the correlation but at this time we will only report the three-variable equation above and move on to another evaluation of the complete mixed list of all the classes listed in the preceding sections.

VII. Summary

The properties dealt with in this chapter are functions of the complex interaction of forces between molecules as well as effects ascribed to interactions within molecules. These effects depend on molecular topology directly and indirectly. The quantitative relationship between structure, expressed as connectivity, and properties is not necessarily linear with $^1\chi$ and, indeed, there may be a complex relationship to connectivity terms.

We have explored the correlation of connectivity with heat of vaporization, boiling point, liquid density, water solubility, and partition coefficient. The form of the connectivity function used in these studies varies from property to property. However, a quality correlation is usually achieved with a reasonable number of variables.

The interaction of molecules in the biological milieu is closely related to the interactions between molecules, as demonstrated in bulk properties. In the next four chapters we investigate various biological activities over a wide range of types.

References

1. G. R. Somayajulu and B. J. Zwolinski, *Trans. Faraday Soc.* **68,** 1971 (1972).
2. L. H. Hall, L. B. Kier, and W. J. Murray, unpublished work.
3. L. H. Hall, L. B. Kier, and W. J. Murray, *J. Pharm. Sci.* **64,** 1974 (1975).
4. L. B. Kier and L. H. Hall, *J. Pharm. Sci.* (in press).
5. L. B. Kier, L. H. Hall, W. J. Murray, and M. Randić, *J. Pharm. Sci.* **65** (1976).
6. L. B. Kier, L. H. Hall, W. J. Murray, and M. Randić, *J. Pharm. Sci.* **64,** 1971 (1975).
7. G. Amidon, S. Yalkowsky, S. Anik, and S. Valvani, *J. Pharm. Sci.* **65** (1976).
8. C. Hansch and J. M. Clayton, *J. Pharm. Sci.* **62,** 1 (1973).
9. W. J. Murray, L. B. Kier, and L. H. Hall, *J. Pharm. Sci.* **64,** 1978 (1975).

Chapter Six

CONNECTIVITY AND NONSPECIFIC BIOLOGICAL ACTIVITY

The preceding chapters have demonstrated quite vividly that structural characteristics influencing a variety of physicochemical properties can be described in terms of molecular connectivity. A set of simply computed, nonempirical numbers, derived from whole numbers reflecting branching complexity, has been shown to correlate very closely with many properties that are additive or constitutive. They correlate very closely with properties that appear to depend on molecular surface area or molecular volume. This latter observation is quite significant in view of the complexities of the molecules considered.

I. Drug–Receptor Phenomena

A. General Principles

It is well established that many drug molecules function in the biological milieu by exerting influences on systems usually called receptors. In addition, the transit of a drug from its portal of entry to this receptor is governed by physicochemical forces comparable to many that we have discussed thus far. The influence of a drug on a receptor is a result of a number of physicochemical phenomena operating in concert among the atoms and functional groups of each.

Fundamentally the interactions comprising a drug–receptor or drug–tissue encounter are electronic phenomena. The total interaction between the two parts may be considered to be a mutual perturbation of the electronic structure of each. In principle it should be possible to describe these events accurately with quantum mechanics. Two major factors bar the way to this approach, at this time. They are, simply, the absence of detailed knowledge about the structure of the receptors, and the enormously complex nature of the probable structures involved in a drug–receptor complex. It is necessary for the present to deal primarily with the structures of drug molecules and to reflect with some form of intuition on what may be occurring in a dynamic sense between the drug and the receptor.

Some efforts have been expended in modeling a receptor and then predicting the course of certain energy changes as drug and receptor models converge [1–4]. A more widely used approach to study drug electronic phenomena has been the estimation of electronic indices of reactivity and the preferred conformations of molecules using various adaptations of molecular orbital theory [5]. These studies have afforded some insight into structural characteristics, at the electronic level, influencing the course of a biological response.

We have shown earlier that molecular connectivity conveys information about molecular structure at an intermediate level, below a quantum mechanical consideration. This information is frequently sufficient to provide a close correlation with many physical properties. It should be quite possible then to study molecular structures and the influence of structural variation on biological response by evaluating a series of drug molecules using concepts of molecular connectivity. This approach is obviously limited to static structures and cannot explicitly consider the electronic influences between atoms. However, molecular connectivity frequently may reflect these influences, where branching and adjacency are major factors governing the quantitative measure of a physicochemical property. It is to be expected that where drug–receptor events are similarly governed, then molecular connectivity may be useful in correlating and predicting biological phenomena.

B. Nonspecific Drug Activity

For the purposes of subclassifying molecular connectivity studies on drug molecules, we will consider in this chapter the possible value of this approach to the study of presumed nonspecific responses. By nonspecific is meant a drug–tissue response that is not governed by the strict structural specificity found in a conventional drug–receptor interaction. The drug molecule does not possess highly prominent functional groups in a strict pattern or constellation, termed a pharmacophore. Instead, the active drug molecules in a series possess seemingly diverse structures. In all probability the molecule is interacting with receptive tissue over a major portion of its surface rather than at selected regions of the molecule. The forces of interaction are probably not concentrated at one or two foci, but are likely diffuse throughout the entire molecule. These forces are thought to be the long-range forces, van der Waals or dispersion interactions. Charge transfer phenomena and hydrogen bonding may also take place. In general, however, the molecule is thought to function as a whole, and structure–activity studies have been concentrated on the complete molecule.

II.   Anesthetic Gases

Current views on the anesthetic gases presume that these molecules function in the fatty phase of nerve tissue, possibly on protein molecules, where they interrupt the normal flow of impulses. The gases are quite varied in their structure and lack, for the most part, what would be considered as receptor-activating functional groups comprising a discrete pharmacophore. Thus the agents are regarded as nonspecific. This attribute is also evident in the wide distribution of effect in the body.

It has been suggested that certain physical properties such as differential solubility in oil and water may relate to the potency of these agents. In a recent study on 31 gases, Hansch *et al.* found a poor correlation between this property and the potency [6]. This property–activity study leaves us in doubt about the structural characteristics that must be present to permit diffusion into a lipid phase near the nerve, or the characteristics necessary to elicit a response at the nerve.

Another assessment of the structure–activity relationship may be possible through the calculation of the molecular connectivity indices for these gases. In Table I are listed 23 of the 31 gases studied by Hansch *et al.* It was necessary to delete from consideration seven gaseous elements since there is no way at this time to consider them from a connectivity point of view. The indices $^1\chi^v$ and $^2\chi^v$ were calculated using the parameters described in Chapter 3. The methyl fluoride molecule proved to be such a bad fit in the correlation with potency that it was not used to obtain the regression equation. The equation and statistics found are

$$\log 1/p = 0.632 \ ^1\chi^v + 0.517 \ ^2\chi^v - 0.501$$

$$r = 0.905, \qquad s = 0.478, \qquad N = 23$$

Individually, each term in χ used in a regression is less satisfactory in relating to the potency ($r = 0.834$ and $r = 0.767$, respectively).

The terms of χ appear to describe structural characteristics that, in concert, contribute in a significant way to the process of nerve anes-

TABLE I

Potency of Anesthetic Gases and Functions of χ

Gas	χ^v	$^2\chi^v$	log $1/p$ Obs[a]	Calc
CF_4	−0.447	−0.150	−1.24	−0.79
C_2F_6	−0.421	−0.485	−1.19	−0.96
SF_6	−0.709	−0.396	−0.75	−1.08
CH_4	0	0	−0.66	−0.44
N_2O	0.349	0.075	−0.18	−0.19
C_2H_4	0.500	0	−0.15	−0.14
C_2H_6	1.000	0	−0.11	0.15
C_3H_8	1.414	0.707	0.05	0.79
C_2H_2	0.333	0	0.15	−0.24
CCl_2F_2	0.980	0.161	0.40	0.23
$CH_3CH:CH_2$	0.986	0.408	0.40	0.37
$(CH_2)_3$	1.500	1.061	0.80	1.04
$CFCl_3$	1.694	1.770	0.82	1.55
CH_3Cl	1.204	0	0.85	0.28
CH_3I	3.430	0	1.15	1.59
CH_3CH_2Cl	1.558	0.851	1.40	0.96
CH_3CH_2Br	2.110	1.403	1.40	1.59
$CH_3CH_2OCH_2CH_3$	1.992	0.781	1.52	1.18
CH_2Cl_2	1.703	1.025	1.52	1.14
CH_3CHCl_2	1.967	2.227	1.59	1.96
$CHCl_3$	2.085	2.510	2.08	2.19
$CF_3CHClBr$	1.794	2.030	2.11	1.75
$CHCl_2CF_2OCH_3$	2.067	0.609	2.66	1.68

[a] From Hansch *et al.* [6].

thesia by these gases. As Hansch *et al.* have shown [6], a relationship to oil and water partition coefficient accounts for only one-third the variance in log $1/p$. A reasonable interpretation of the results in Table I is that the terms of χ are reflecting structural characteristics influencing the interaction of the molecule with portions of the nerve cell surface. This brings to mind long-range forces such as dispersion interaction.

Hansch *et al.* have suggested that a potential hydrogen bonding between the gas and the receptor tissue may result in the observed anesthesia. By introducing a dummy parameter in their data [6] they were able to achieve a significant improvement in the correlation (from r = 0.613 to r = 0.947). If a similar treatment is given to the molecules in

Table I, namely, the addition of a dummy parameter to reflect the possibility of hydrogen bonding ($H = 1$) or the absence of this potential ($H = 0$), the correlation is not greatly improved ($r = 0.928$). This does not refute the possible role of hydrogen bond interactions in the case of some of the gases, but it does suggest that this possibility is subsumed for the most part in the structural description encoded in the χ terms.

III.   Nonspecific Local Anesthetic Activity

Another study illustrates the role of molecular connectivity in describing structural features influencing nonspecific drug action. Agin *et al.* [7] have published their findings on the nerve-blocking concentration of over 30 widely varying chemical structures. They observed a good correlation between the log minimum blocking concentration (log MBC) and an approximation of the London or dispersion interaction energy E_L between a molecule and a conducting surface. This energy was deduced from the approximation $E_L = \alpha I/8r^3$ [8], where α is the electronic polarizability of the molecule, I the ionization potential, and r the interacting distance. The values for I were eclectic values from the literature, Huckel MO approximations, and extrapolated estimates.

These authors have emphasized in their conclusions that the molecules studied for their local anesthetic potency exhibited a marked lack of structural specificity. They concluded that the significant feature contributing to anesthetic potency was the participation of each molecule in a dispersion interaction with some portion of nervous tissue, thereby interrupting its conduction.

These same molecules have been recently examined using molecular connectivity to describe their structures [9] (Table II). The simple $^1\chi$ term was found to correlate closely with the log MBC value:

$$\log \text{MBC} = 3.60 - 0.779 \, ^1\chi, \qquad r = 0.982, \quad s = 0.409, \quad N = 36$$

In this case the addition of the valence derived $^1\chi^v$ term did not improve the quality of the correlation.

It is interesting to note that the χ term alone correlated as well with the log MBC values as did the α and I values of Agin *et al.* One explanation may lie in the probable poor quality of the I values used in the study. The second explanation lies in the observation that the I values for the molecules in Table II, used by Agin *et al.*, fall in a narrow range, approximately 2 eV.

If the latter explanation is considered operative, then the principal

TABLE II

Local Anesthetic Activity, Polarizability, and χ

	α		log MBC	
Anesthetic	Obs	Calc	Obs[a]	Calc
Methanol	8.2	10.9	3.09	2.79
Ethanol	12.9	14.7	2.75	2.47
Acetone	16.2	17.7	2.6	2.23
2-Propanol	17.6	17.7	2.55	2.23
Propanol	17.5	19.3	2.4	2.09
Urethane	23.2	27.3	2.0	1.44
Ether	22.5	24.0	1.93	1.71
Butanol	22.1	24.0	1.78	1.71
Pyridine	24.1	24.8	1.77	1.65
Hydroquinone	29.4	32.1	1.4	1.05
Aniline	31.6	28.4	1.3	1.35
Benzyl Alcohol	32.5	33.4	1.3	0.935
Pentanol	26.8	28.6	1.2	1.33
Phenol	17.8	28.4	1.0	1.35
Toluene	31.1	28.4	1.0	1.35
Benzimidazole	40.2	33.7	0.81	0.901
Hexanol	31.4	33.2	0.56	0.949
Nitrobenzene	32.5	36.8	0.47	0.651
Quinoline	42.1	38.3	0.3	0.528
8-Hydroxyquinoline	44.7	42.6	0.3	0.174
Heptanol	36.0	37.9	0.2	0.567
2-Naphthol	45.4	42.0	0.0	0.228
Methyl anthranilate	48.9	45.6	0.0	−0.072
Octanol	40.6	42.5	−0.16	0.186
Thymol	47.3	44.3	−0.52	0.052
o-Phenanthroline	57.8	52.1	−0.8	−0.602
Ephedrine	50.2	50.3	−0.8	−0.453
Procaine	67.0	72.6	−1.67	−2.29
Lidocaine	72.5	72.8	−1.96	−2.31
Diphenhydramine	79.5	78.2	−2.8	−2.75
Tetracaine	79.7	81.5	−2.9	−3.03
Phenyltoloxamine	79.9	78.1	−3.2	−2.74
Quinine	93.8	91.5	−3.6	−3.85
Physostigmine	82.4	75.4	−3.66	−2.52
Caramiphen	87.0	87.0	−4.0	−3.48
Dibucaine	103.6	105.2	−4.2	−4.97

[a] From Agin *et al.* [7].

influence in Agin's study would be the magnitude of the polarizability α. It would be expected then that χ for this series should correlate closely with the α values. This was indeed found to be true [9]. The relationship for the compounds in Table II between χ and α is

$$\alpha = 9.503\chi + 0.840, \qquad r = 0.993, \quad s = 3.108, \quad N = 36$$

This study presents a contrast to the study of anesthetic gases in Section II. The χ term is sufficient to describe the structural characteristics relating to local anesthetics in Table II. In the list of gases in Table I, terms derived from the valence of molecular vertices are necessary to describe adequately the salient structural features. A possible explanation is that the gases are much richer in noncarbon atoms, relative to the total number of atoms in the molecules. As a consequence, it is necessary to account for their more dominant influence in the molecule by considering the heteroatom valencies more explicitly. Thus, χ^v for these gases is necessary for an adequate description.

In contrast, the molecules in Table II are largely carbon containing. They are also largely planar or aromatic molecules. In this case, the χ term alone is sufficient for a useful structural description. These differences should be monitored as more studies are performed in order to anticipate the connectivity terms most useful in specific instances.

IV.   Nonspecific Narcotic Activity

The two preceding studies have dealt with structurally nonspecific molecules presumably acting at selected tissues. Another category of nonspecific biological activity that has been studied for structure–activity is whole animal narcosis [10]. These studies, which should accurately be called property–activity studies, have attempted to relate the narcotic influence on larvae or tadpoles to physical properties such as partition coefficient or polarizability.

Several aspects of this type of study are worth comment prior to a consideration of the role that connectivity may play in structure–activity studies. The quality of the data is probably limited to 5 to 10% error in measurement. Thus any correlation with these data exceeding $r = 0.95$ may not be highly meaningful, since the variance in r^2 may be entirely due to data error.

A second characteristic of this kind of data is the possibility that several mechanisms of action leading to the narcotic effect may be operating. It cannot be assumed, in every case, that the rate-limiting step leading to narcosis, for example, is the passage of the drug into the organism. Thus, this single interpretation of a partition coefficient correlation may not adequately reflect reality. The correlation with partition coefficient may actually be reflecting several mechanisms involving subclasses of drugs in a list, each giving a roughly comparable correlation. With the partition coefficient it is not possible to do much to factor out subclasses of structural influences in several mechanistic roles, since the physical property is not a fundamental structural description.

This opportunity may be available in studying the connectivity relationship to narcotic effect. Two studies explore this possibility.

A. Narcosis of *Arenicola* Larvae

In Table III are listed the narcotic concentration, pC, of 20 varied compounds [11], against the *Arenicola* larvae. Calculations of $^1\chi$ and $^1\chi^v$

TABLE III

Narcosis of Arenicola Larvae and χ

Molecule	χ^v	pC Obs[a]	pC Calc	Residual
Methanol	0.447	−0.40	−0.23	−0.17
Ethanol	1.023	−0.01	0.32	−0.33
Propanol	1.523	0.47	0.80	−0.33
Isopropanol	1.413	0.41	0.70	−0.29
Butanol	2.023	1.06	1.27	−0.21
Pentanol	2.523	1.64	1.76	−0.12
Octanol	4.023	3.00	3.21	−0.21
Ethyl acetate	1.904	0.89	1.16	−0.27
Ethyl propionate	2.465	1.41	1.70	−0.29
Ethyl butyrate	2.965	1.89	2.18	−0.30
Ethyl nitrate	1.496	1.29	0.62	0.67
Ethyl valerate	3.465	3.00	2.67	0.33
Methyl carbamate	1.105	0.18	0.40	−0.22
Ethyl carbamate	1.693	0.47	0.95	−0.50
Benzene	2.000	2.28	1.25	1.03
Toluene	2.410	2.29	1.65	0.64
Nitromethane	0.742	0.36	0.05	0.31
Acetonitrile	0.727	0.02	0.04	−0.02
Carbon tetrachloride	2.408	2.59	1.65	0.94
Acetanilide	3.114	1.68	2.33	−0.65

[a] From Lillie [11].

TABLE IV

Narcosis of Arenicola Larvae and χ: Subclass One from Table III

Molecule	pC Obs[a]	pC Calc	Molecule	pC Obs[a]	pC Calc
Methanol	−0.40	−0.49	Ethyl propionate	1.41	1.47
Ethanol	−0.01	0.06	Ethyl butyrate	1.89	1.96
Propanol	0.47	0.55	Ethyl valerate	3.00	2.45
Isopropanol	0.41	0.44	Methyl carbamate	0.18	0.15
Butanol	1.06	1.04	Ethyl carbamate	0.47	0.72
Pentanol	1.64	1.53	Acetonitrile	0.02	−0.22
Octanol	3.00	3.00	Acetanilide	1.68	2.10
Ethyl acetate	0.89	0.92			

[a] From Lillie [11].

for each molecule are made using parameters described in Chapter 3. The $^1\chi$ term shows a poor correlation, while the $^1\chi^v$ term exhibits a fair correlation:

$$pC = 0.969 \; ^1\chi^v - 0.687, \qquad r = 0.896, \quad s = 0.483, \quad N = 20$$

The calculated values and residuals are also listed in Table III. A close inspection of the residuals reveals that most of the compounds are predicted by the equation to have a high value of pC, though by only about one-half the standard deviation. The remaining compounds are calculated to have a significantly lower pC value, based on prediction from their $^1\chi^v$ values. A suspicion is aroused that there may indeed be at least two equations subsumed into the one above. This is reinforced by an inspection of the molecules that are predicted to have modestly high pC values. They include the alcohols, ethers, carbamates, a nitrile, and an amide. In contrast, the compounds predicted to be quite a bit below the equation line are two alkyl nitrates, benzene, toluene, and carbon tetrachloride.

Individual regression analyses of the $^1\chi^v$ values for each group against the pC values produces informative results. The equation and statistics for the 15-member group are

$$pC = 0.978 \; ^1\chi^v - 0.935, \qquad r = 0.978, \quad s = 0.225, \quad N = 15$$

TABLE V

Narcosis of Arenicola Larvae and χ: Subclass Two from Table III

| | pC | |
Molecule	Obs[a]	Calc
Ethyl nitrate	1.29	1.20
Benzene	2.28	2.04
Toluene	2.29	2.55
Nitromethane	0.36	0.47
Carbon terachloride	2.59	2.54

[a] From Lillie [11].

The observed and predicted values are reproduced in Table IV. The results of the correlation between pC and the remaining five compounds are

$$pC = 1.283 \ ^1\chi^v - 0.561, \qquad r = 0.979, \quad s = 0.218, \quad N = 5$$

These results are shown in Table V. Figure 1 shows the two equation lines.

These results support the prediction that the narcotic results in this

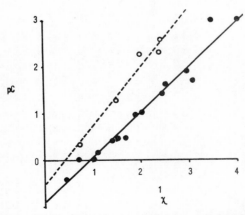

Fig. 1. Narcotic concentration pC of compounds versus *Arenicola* larvae. Solid line is equation from Table IV and dashed line is equation from Table V.

TABLE VI

Tadpole Narcosis and χ

Molecule	$^1\chi$	$^1\chi^v$	$^1\chi - {^1\chi^v}$	log 1/c Obs[a]	log 1/c Calc
Methanol	1.000	0.447	0.55	0.24	0.19
Ethanol	1.414	1.023	0.39	0.54	0.74
Propanol	1.914	1.523	0.39	0.96	1.28
Butanol	2.414	2.023	0.39	1.42	1.80
Octanol	4.414	4.023	0.39	3.40	3.88
Isopropanol	1.732	1.413	0.32	0.89	1.16
Isobutanol	2.270	1.879	0.39	1.35	1.65
tert-Butanol	2.000	1.724	0.28	0.89	1.49
Isoamyl alcohol	2.770	2.379	0.39	1.64	2.17
tert-Amyl alcohol	2.561	2.284	0.28	1.24	2.07
1,3-Dichloro-2-propanol	2.808	2.777	0.03	1.92	2.58
Thymol	5.109	3.905	1.20	4.26	3.76
1,3-Dimethoxybenzene	4.864	3.046	1.82	3.35	2.86
1,4-Dimethoxybenzene	4.864	3.046	1.82	3.05	2.86
Acetone	1.732	1.204	0.53	0.54	0.95
Butanone	2.270	1.765	0.21	1.04	1.53
3-Pentanone	2.808	2.325	0.48	1.54	2.11
2-Pentanone	2.770	2.265	0.21	1.72	2.05
Acetophenone	4.305	2.865	1.44	3.03	2.67
Acetal	3.808	3.040	0.77	1.98	2.86
Ethyl ether	2.414	1.992	0.42	1.57	1.77
Anisole	3.932	2.532	1.40	2.82	2.32
Methyl acetate	2.270	1.316	0.95	1.10	1.06
Ethyl formate	2.414	1.467	0.95	1.15	1.15
Ethyl acetate	2.770	1.904	0.87	1.52	1.67
Ethyl propionate	3.308	2.465	0.84	1.96	2.26
Propyl acetate	3.270	2.405	0.87	1.96	2.20
Ethyl butyrate	3.808	2.965	0.84	2.37	2.76
Ethyl isobutyrate	3.681	2.847	0.83	2.24	2.64
Butyl acetate	3.770	2.905	0.87	2.30	2.71
Isobutyl acetate	3.626	2.760	0.87	2.24	2.56
Ethyl valerate	4.308	3.465	0.84	2.72	3.30
Amyl acetate	4.270	3.405	0.87	2.72	3.23
Butyl valerate	5.308	4.465	0.84	3.60	4.34
Methyl carbamate	2.270	1.105	1.17	0.57	0.84
Ethyl carbamate	2.770	1.693	1.08	1.39	1.45
Phenyl carbamate	4:788	2.812	1.98	3.19	2.62
Pentane	2.414	2.414	0.0	2.55	2.20
Pentene	2.414	2.024	0.39	2.65	1.80
Benzene	3.000	2.000	1.000	2.68	1.77
Xylene	3.805	2.827	0.98	3.42	2.64

TABLE VI—Continued

Molecule	$^1\chi$	$^1\chi^v$	$^1\chi - {}^1\chi^v$	log $1/c$ Obs[a]	log $1/c$ Calc
Naphthalene	4.966	3.405	1.56	4.19	3.23
Phenanthrene	6.949	4.815	2.13	5.43	4.70
Ethyl chloride	1.414	1.588	−0.17	2.35	1.31
Ethyl bromide	1.414	2.110	−0.70	2.57	1.89
Ethyl iodide	1.414	3.132	−1.72	2.96	2.95
Ethylene chloride	1.914	2.203	−0.29	2.64	1.98
Chloroform	1.732	2.085	−0.35	2.85	1.86
Nitromethane	1.732	0.742	0.90	1.09	0.46
Acetonitrile	1.414	0.724	0.69	0.44	0.45
Azobenzene	6.949	4.468	2.48	4.74	4.34
Acetaldehyde oxime	1.914	1.036	1.40	0.92	0.77

[a] From Overton [12].

study may well involve two distinct mechanisms of action, each structurally nonspecific, in a broad sense, but still individual to a noticeable degree.

The subclassification into two small groups, 5 and 15 compounds, may not be sufficiently compelling evidence to accept the conclusion of two operating mechanisms. A larger group of diverse structures is available in a similar study to test this possibility.

B.   Tadpole Narcosis

At the turn of the century, Overton found that a large number of compounds of widely diverse chemical structure exhibited varying degrees of narcosis on frog tadpoles [12] (Table VI). Leo *et al*. [10] have found a good correlation between the effective concentration, log $1/c$, and the partition coefficients of the molecules in octanol and water ($r = 0.955$, $s = 0.343$).

This study presents an opportunity to explore the possible relationship between molecular connectivity and a nonspecific biological activity, using a larger set of compounds than in the previous sample.

We have found that $^1\chi^v$ gives a fair correlation with activity. It was necessary to delete diethyl tartrate from the regression analysis since this molecule was predicted to be very much more active than the

TABLE VII

Tadpole Narcosis and χ: Subclass One from Table VI

	log $1/c$	
Molecule	Obs[a]	Calc
Methanol	0.24	0.13
Ethanol	0.54	0.62
Propanol	0.96	1.04
Butanol	1.42	1.47
Octanol	3.40	3.17
Isopropanol	0.89	0.95
Isobutanol	1.35	1.35
tert-Butanol	0.89	1.21
Isoamyl alcohol	1.64	1.77
tert-Amyl alcohol	1.24	1.69
1,3-Dichloro-2-propanol	1.92	2.11
Acetone	0.54	0.77
Butanone	1.04	1.25
3-Pentanone	1.54	1.73
2-Pentanone	1.72	1.68
Acetal	1.98	2.34
Ethyl ether	1.57	1.44
Methyl acetate	1.10	0.87
Ethyl formate	1.15	0.94
Ethyl acetate	1.52	1.37
Ethyl propionate	1.96	1.85
Propyl acetate	1.96	1.80
Ethyl butyrate	2.37	2.26
Ethyl isobutyrate	2.24	2.16
Butyl acetate	2.30	2.21
Isobutyl acetate	2.24	2.10
Ethyl valerate	2.72	2.69
Amyl acetate	2.72	2.64
Butyl valerate	3.60	3.55
Methyl carbamate	0.57	0.69
Ethyl carbamate	1.39	1.19
Acetonitrile	0.44	0.36
Acetaldehyde oxime	0.93	0.63

[a] From Overton [12].

experimental data. The predicted values for 52 compounds are shown in Table VI as derived from the regression equation:

$$\log 1/c = 1.041 \, ^1\chi^v - 0.308, \qquad r = 0.886, \quad s = 0.540, \quad N = 52$$

By applying the same analysis of the residuals for this list as described in the previous study, it can be seen that 33 of the compounds predict a high value for $\log 1/c$. These molecules are varied in structure but are all nonaromatic esters, ketones, alcohols, or nitrogen-containing compounds. The remaining compounds have $^1\chi^v$ values that predict a lower $\log 1/c$ value. These molecules are aromatic, aliphatic hydrocarbons or alkyl halides.

An analysis of the subset containing the 33 compounds reveals a good correlation:

$$\log 1/c = 0.855 \, ^1\chi^v - 0.259, \qquad r = 0.975, \quad s = 0.187, \quad N = 33$$

The fitted values from this equation are shown in Table VII. This close correlation suggests that the $^1\chi^v$ values for this list of compounds are describing salient structural features related to the activity. A further

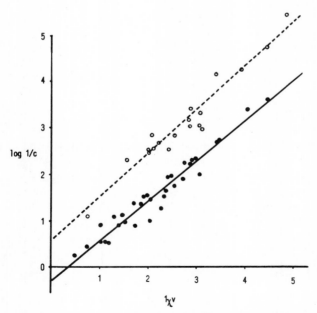

Fig. 2. Narcotic concentration $\log 1/c$ of compounds versus tadpoles. Solid line is equation from Table VII and dashed line is equation from Table VIII.

TABLE VIII

Tadpole Narcosis and χ: Subclass Two from Table VI

Molecule	log $1/c$	
	Obs[a]	Calc
Thymol	4.26	4.26
1,3-Dimethoxybenzene	3.35	3.44
1,4-Dimethoxybenzene	3.05	3.44
Acetophenone	3.03	3.27
Anisole	2.82	2.95
Phenyl carbamate	3.19	3.22
Pentane	2.55	2.84
Pentene	2.65	2.47
Benzene	2.68	2.45
Xylene	3.42	3.23
Naphthalene	4.19	3.78
Phenanthrene	5.43	5.12
Ethyl chloride	2.35	2.03
Ethyl bromide	2.57	2.55
Ethyl iodide	2.96	3.52
Ethylene chloride	2.64	2.64
Chloroform	2.85	2.53
Nitromethane	1.09	1.25
Azobenzene	4.74	4.79

[a] From Overton [12].

interpretation is that a single mechanism may be operating at the tissue or cellular level in which these compounds are participating.

It is necessary to establish whether the remaining molecules from the list may be acting through a separate mechanism or whether they are random, poorly fitting points in the regression. These 19 compounds appear to exhibit a relationship between $^1\chi^v$ and log $1/c$ that is almost as good as the larger subset of compounds. This second subset, containing 19 molecules, is shown in Table VIII. The equation and statistics are

$$\log 1/c = 0.949 \; {}^1\chi^v + 0.553, \qquad r = 0.962, \quad s = 0.272, \quad N = 19$$

Clearly, the $^1\chi^v$ value collects these molecules into a closely related list relative to the action on tadpoles.

These analyses support a postulate of at least two mechanisms of action operating in this list. The equation lines are shown in Fig. 2.

The two subsets of compounds in Tables VII and VIII exhibit different characteristics in a connectivity sense. The 33 compounds in Table VII are calculated to have each a $^1\chi$ value averaging close to 0.6 more than the corresponding $^1\chi^v$ value. This differential, which we can provisionally call a valence–connectivity differential (VCD), and which is shown in Table VI as $^1\chi - {}^1\chi^v$, may be considered to reflect a common structural characteristic that may be influential in the biological action.

The second subset in Table VIII reveals a wider diversity in the VCD values. The eleven aromatic molecules have VCD values averaging 1.6, the alkane and alkene VCD values are zero or smaller, while the alkyl halides have significant negative values. These observations suggest the possibility of further subclassification into separate mechanisms for this list of molecules.

V. The Question of Nonspecific Drug Action

A broader question is raised by these last two studies. Does the term "nonspecific" really convey an accurate impression of the drug–tissue response? By the factoring process we have demonstrated a significant refinement in the relationship of connectivity to biological activity within subclasses of the original list of molecules. This suggests that the procedure has revealed some measure of identity within each subclass different from the other compounds in the list. The further subdivision of the list on the basis of connectivity differences now surfaces as a distinct possibility. The reality of several mechanisms of action leading to a common measured response, each with a characteristic SAR, is a profile of structurally specific drug–tissue or drug–receptor interactions. Each subclass of compounds in a list, such as those just described, must possess a commonality in the form of similar structural features that are important in the interaction phenomena. This is structure specificity.

We conclude this chapter by returning to our original definition of nonspecific drug action. Can a drug act on a tissue or cell in a structurally nonspecific manner at the molecular level apart from a strictly bulk effect? The analyses presented here raise some doubt. Certainly the physicochemical events preceding a receptive tissue encounter may be nonspecific structurally. These would include membrane passage by passive mechanisms as well as fat tissue solubility. But where an intermolecular encounter is involved at receptive tissue, it seems probable that some definitive characteristics uniting a set of molecules structurally must be involved. This common set of features may be appreciable aromatic surface area or alkyl chain, or an electron-rich center as found on a heteroatom. These are, however, specific enough so that other molecules with grossly similar physical properties may be excluded from the particular mechanism if they lack the salient structures.

The studies in this chapter provide examples of the kind of relationships that can be developed between structure and biological activity. They also illustrate how molecular connectivity can be used to analyze these relationships and to gain further insight into structural similarities, important in a large group of molecules. The ability of χ as a molecular structure descriptor is certainly attested to by the ability of this index to organize and rank molecules of very diverse structure, as found among the anesthetic gases. Subsequent chapters will reveal other examples of SAR studies and opportunities for new insight.

References

1. L. B. Kier and H. S. Aldrich, *J. Theor. Biol.* **46**, 529 (1974).
2. H. D. Holtje and L. B. Kier, *J. Theor. Biol.* **48**, 197 (1974).
3. L. B. Kier and H. D. Holtje, *J. Theor. Biol.* **49**, 401 (1975).
4. L. H. Hall and L. B. Kier, *J. Theor. Biol.* **58**, 177 (1976).
5. L. B. Kier, "Molecular Orbital Theory in Drug Research." Academic Press, New York, 1971.
6. C. Hansch, A. Vittoria, C. Silipo, and P. Jow, *J. Med. Chem.* **18**, 546 (1975).
7. D. Agin, L. Hersh, and D. Holtzman, *Proc. Nat. Acad. Sci.* **53**, 952 (1965).
8. H. Casimer and D. Polder, *Phys. Rev.* **73**, 360 (1948).
9. L. B. Kier, L. H. Hall, W. J. Murray, and M. Randić, *J. Pharm. Sci.* **64**, 1971 (1975).
10. A. Leo, C. Hansch, and C. Church, *J. Med. Chem.* **12**, 766 (1969).
11. R. S. Lillie, *J. Physiol.* **31**, 255 (1913).
12. E. Overton, "Studies on Narcosis." Fischer, Jena, Germany, 1901.

Chapter Seven

SUBSTITUENT GROUP STRUCTURE–ACTIVITY RELATIONSHIPS

It has been demonstrated many times that the variation of the structures of molecules in a series will influence, in a quantitative way, the range of biological responses. At the molecular level several interpretations may be placed on this observation. The impact of changing a portion of the molecule, usually referred to as a substituent, may influence thermodynamic properties so that the relative accessibility of the drug to the receptor is modified. This accessibility may be a function of the relative ease of absorption from the gut or through a cell wall, or it may depend on partitioning between an aqueous and a lipid interface. The structural variation thus introduced is indirectly responsible for a measured biological response in the series. The parameters of structure used to relate to response in this case are indirectly correlated with drug–receptor events. They measure the influence on some intervening limiting event.

A second explanation for the perceived influence of substituent group variation on activity is the direct influence of the modified substituent on the drug–receptor interaction. In this case, the substituent may be thought of as playing a significant role in the perturbation process at the receptor.

It has become a common practice to attempt to relate structural variation and drug response by measuring some intermediate physical property that may relate to the activity. Norrington *et al.* refer to this as a physical property–activity relationship (PAR) and not a true structure–activity (SAR) relationship [1]. One commonly used property is the partition coefficient between oil and water. The information from the comparison of parallel properties, such as partition coefficient and biological activity, is useful in extrapolations, where the property can be measured easily or, less reliably, predicted.

Reference to a molecule by a physical property such as partition coefficient is not a description of the structure of the molecule. The insight gained is limited to the realm of structural variations that can be measured or predicted for that physical property. Further, the interpretation is a forced one in view of the nature of the property. Thus the only meaningful interpretation of the relationship between partition coefficients in a series of molecules and some biological response is that the partitioning process is influencing the response at some limiting step. Attempts to extend the interpretation of these PAR studies to invoke secondary binding influences involving the substituent at the receptor are not on firm ground.

A more fundamental approach to describing a drug molecule would carry with it a broader ability to convey structural information, to extrapolate to new structures, and to permit a wider latitude of mechanistic interpretation. In this chapter we will examine the role that molecular connectivity may play in assessing structure as it influences biological events. In particular, we will focus on the structure–activity relationships that may be revealed in studies of substituent groups on active molecules.

I. Alcohol Narcosis of Barnacle Larvae

We have pointed out in the preceding chapter that narcosis of simple organisms like larvae or tadpoles may not be an actual nonspecific effect. It is quite likely that some rather definite structure-dependent processes are extant, which lead to observed gradations of response. If this is true, then structure modification of a substituent in molecules of a series leaves intact a portion of the molecule responsible for the specific effect. Modification of the substituent may influence the arrival of the drug to the receptive tissue or it may directly influence the interaction.

The activity of several alcohols in narcotizing barnacle larvae [2] provides an interesting study of the role molecular connectivity may play in SAR. The test results are shown in Table I. Calculations of $^1\chi$ and the activity have been reported [3] and are included in the table. The relationship is

$$pC = 1.043 \ ^1\chi - 1.123, \qquad r = 0.987, \quad s = 0.163, \quad N = 15$$

A similar analysis using the partition coefficient gave the same r and s values [4].

TABLE I

Barnacle Larvae Narcosis and χ

| | | pC | |
Molecule	$^1\chi$	Obs[a]	Calc
Methanol	1.000	−0.14	−0.08
Ethanol	1.414	0.28	0.35
Propanol	1.914	0.79	0.87
Isopropanol	1.732	0.92	0.68
Butanol	2.414	1.46	1.39
Pentanol	2.914	1.84	1.91
Hexanol	3.414	2.41	2.43
Heptanol	3.914	3.02	2.96
Octanol	4.414	3.62	3.48
2-Methyl-1-propanol	2.270	1.54	1.24
2-Butanol	2.270	1.16	1.24
2-Methyl-2-propanol	2.000	0.98	0.96
3-Methyl-1-butanol	2.770	1.86	1.77
2-Methyl-2-butanol	2.561	1.34	1.54
Benzyl Alcohol	3.432	2.15	2.45

[a] From Crisp and Marr [2].

A correlation between pC and $^1\chi^v$ and between pC and $^1\chi$ with $^1\chi^v$ were quite good, but inferior to $^1\chi$ alone. From this we can infer that $^1\chi$ supplies a very adequate description of the set of molecules in reference to this biological activity. The fact that $^1\chi$ and $^1\chi^v$ give as good a correlation suggests a constant role for the hydroxyl group in the mechanism of action. Such a conclusion is based on the fact that it is not necessary in the description of the molecules to use the valence deltas. Thus its position in the molecule as a primary, secondary, or tertiary hydroxyl group is adequately depicted by $^1\chi$. If $^1\chi^v$ were a superior index, it would suggest that this positional difference was playing a more prominent role in influencing the activity. A possible conclusion is that the hydroxyl group is involved in donating a hydrogen in a hydrogen bond at a receptor. This phenomenon would not be greatly dependent on the class of hydroxyl group.

II.   Cytochrome Conversion by Phenols

Ichikawa and Yamano have reported on the conversion of cytochrome P-450 to P-420 in the rabbit liver [5]. The concentration for half-

TABLE II

Cytochrome P-450 Conversions by Substituted Phenols

| Phenol substituent | $^1\chi^v$ | pC | |
		Obs[a]	Calc
H	2.134	1.07	0.95
3-Hydroxy	2.268	0.81	1.06
3-Amino	2.333	0.46	1.11
4-Methyl	2.545	1.48	1.29
4-Carboxy	2.722	1.15	1.43
3-Methyl	2.545	1.50	1.29
2-Chloro	2.653	1.60	1.38
3-Ethyl	3.105	1.82	1.74
4-Bromo	3.037	2.04	1.69
2-Iodo	3.766	2.09	2.28
2,4-Dichloro	3.165	2.11	1.79
2,4,6-Trichloro	3.604	2.21	2.15
2,3,4,6-Tetrachloro	4.208	2.65	2.64
Pentachloro	4.734	2.90	3.07

[a] From Ichikawa and Yamano [5].

conversion, pC, was measured for a series of substituted phenols. An analysis of these molecules using the connectivity indices reveals a correlation between $^1\chi^v$ and the concentration to produce the transformation, pC, shown in Table II. The relationship is

$$pC = 0.816\ ^1\chi^v - 0.789, \qquad r = 0.914, \quad s = 0.291, \quad N = 14$$

An analysis using $^1\chi$ leads to a much poorer correlation, $r = 0.675$, while a relationship between pC and $^1\chi$ with $^1\chi^v$ leads to no improvement over the relationship expressed for $^1\chi$. The possible electronic influence of the substituent group on the ring or phenolic hydroxyl group has been considered. The Hammett sigma constant for the substituents does not correlate as well as $^1\chi^v$ nor does it improve $^1\chi^v$ when considered in a multiple regression.

These results suggest the possibility that the substituents may not be exclusively involved at a receptor but may be participating, along with the benzene ring, in a dispersion-type interaction with a receptor feature. This reinforces the concept of a primary binding through the phenolic hydroxyl group that is imparting the qualitative response.

III. Enzyme Inhibitors

A number of enzyme inhibitor studies are available in suitable form to permit an examination of the possible relationship that a connectivity index may have on the activity. Kier *et al*. have reported several such studies [6].

A. Thymidine Phosphorylase Inhibitors

A series of eleven alkyl and arylalkyl derivatives of thymidine (N_1-substituted) have been made and tested for potency in terms of 50% inhibitory concentration [7]. These results are shown in Table III. The relationship to connectivity is

$$\log 1/c = 0.366 \,^1\chi - 3.364, \qquad r = 0.920, \quad s = 0.213, \quad N = 11$$

This example is an uncomplicated case of a variable substituent being positioned in only one location in molecules in a series. It would be possible under these circumstances to calculate the $^1\chi$ value for only the substituent and the nonvarying atom at the point of attachment. This

TABLE III

Inhibitors of Thymidine Phosphorylase

R	$^1\chi$	log 1/c Obs[a]	Calc
Methyl	3.698	−2.30	−2.01
Butyl	5.236	−1.35	−1.44
Isopentyl	5.592	−1.30	−1.31
Cyclopentyl	5.270	−1.28	−1.43
Isohexyl	6.092	−1.17	−1.13
Pentyl	5.736	−1.15	−1.26
3-Phenyl propyl	7.254	−1.11	−0.71
2-Phenyl ethyl	6.754	−0.80	−0.89
Phenyl methyl	6.254	−0.76	−1.07
4-Phenyl butyl	7.754	−0.60	−0.52
5-Phenyl pentyl	8.254	−0.32	−0.34

[a] From Baker and Kazawa [7].

TABLE IV

Inhibitors of Adenosine Deaminase

R	$^1\chi^v$	$\log(I/S)_{0.5}$ Obs[a]	$\log(I/S)_{0.5}$ Calc
Methyl	3.163	−0.86	−0.98
Ethyl	3.739	−0.79	−0.73
Propyl	4.239	−0.52	−0.51
Butyl	4.739	−0.36	−0.28
Pentyl	5.239	−0.15	−0.06
Hexyl	5.739	0.15	0.16
Heptyl	6.239	0.49	0.38
Octyl	6.739	0.62	0.61

[a] From Schaeffer *et al.* [8].

procedure is valid except in the case of a hydrogen atom as a substituent. Thus, if the ring nitrogen atom is assigned a δ value of 3, the $^1\chi$ values for the first four substituents in Table III are 0.577, 2.115, 2.471, and 3.150. These values are expected to correlate with the activity equally well as the $^1\chi$ value for the entire molecule. The case of the hydrogen on the nitrogen has to be treated differently. The δ value on the nitrogen becomes 2. Therefore the $^1\chi$ corresponding to the hydrogen substituent is given as

$$[(3 \times 2)^{-1/2} + (2 \times 2)^{-1/2}] - [(3 \times 3)^{-1/2} + (2 \times 3)^{-1/2}] = 0.166$$

Group values of $^1\chi^v$ could be calculated in a similar manner. Again, the value for a hydrogen substituent is treated as described above in view of the change of valence for the nitrogen atom.

The calculated group values for $^2\chi$ and beyond present a greater complication. The $^2\chi$ value for a methyl substituent includes δ values for two ring atoms flanking the nitrogen. Values for $^3\chi$ terms embrace three atoms of the ring. These cases are best dealt with using the entire molecule to compute $^m\chi_t$ terms. An additional consideration is the possible use of the χ value for an entire molecule for comparison with similar ring systems or the possibility of ring substitution at another position.

B. Adenosine Deaminase Inhibitors

A series of adenosine derivatives have been reported to inhibit the enzyme adenosine deaminase [8]. The results of a connectivity analysis are shown in Table IV. The relationship of χ to inhibitory potency is

$$\log(I/S)_{0.5} = 0.445\ {}^1\chi - 2.395, \qquad r = 0.990, \quad s = 0.086, \quad N = 8$$

This case is not a severe test for the connectivity index since the increment between homologs is exactly 0.5 after the ethyl derivative. Nevertheless, the very high correlation and low standard error indicate the meaningfulness of ${}^1\chi$ in this case.

C. Butyrylcholinesterase Inhibitors

A series of piperidine carboxylic acid amide derivatives have been found to inhibit the enzyme butyrylcholinesterase [9]. The structural modifications are both mono- and dialkyl substitutions of the amide nitrogen. This is a somewhat different case than a simple monosubstituent since the δ for the nitrogen atom assumes values of 1, 2, or 3 in the series. The correlation with ${}^1\chi$ is quite good, as shown in Table V, with

TABLE V

Butyrylcholinesterase Inhibitory Potency

			pI$_{50}$	
R^1	R^2	${}^1\chi$	Obs[a]	Calc
H	H	8.736	4.21	4.16
H	Methyl	9.274	4.46	4.49
H	Ethyl	9.774	4.86	4.80
Methyl	Methyl	9.647	4.66	4.72
Methyl	Ethyl	10.147	5.01	5.02
Ethyl	Ethyl	10.647	5.28	5.33
Propyl	Propyl	11.647	5.98	5.94

[a] From Clayton and Purcell [9].

the equation

$$pI_{50} = 0.611 \,^1\chi - 1.173, \qquad r = 0.996, \quad s = 0.055, \quad N = 7$$

The use of group χ values in this case would require a specific consideration of the changes in the δ for the nitrogen in the series.

IV.　Microbial Inhibition

Another example of the use of χ to study molecules substituted in more than one position is the inhibition of *M. tuberculosis* by a series of alkyl bromophenols [10], as shown in Table VI. Both $^1\chi$ and $^1\chi^v$ give good correlations with the inhibitory concentration pC, but $^1\chi^v$ is clearly superior. The relationship is

$$pC = 0.792 \,^1\chi^v - 0.543, \qquad r = 0.986, \quad s = 0.176, \quad N = 14$$

Because of the variation in the substitution patterns in this series (1; 1,4; 1,2,4; 1,2,3,4,5) it is obviously impossible to deal with the compounds other than by calculating the χ indices for the entire molecule.

TABLE VI

Inhibition of M. Tuberculosis by Substituted Phenols and χ

		pC	
Phenol substituent	$^1\chi^v$	Obs[a]	Calc
H	2.134	0.95	1.15
4-Bromo	3.037	1.94	1.86
2-Methyl, 4-bromo	3.454	2.35	2.19
2-Ethyl, 4-bromo	4.015	2.70	2.64
2-Propyl, 4-bromo	4.515	3.18	3.03
2-Butyl, 4-bromo	5.015	3.76	3.43
2-Pentyl, 4-bromo	5.515	3.99	3.83
2-sec-Pentyl, 4-bromo	5.435	3.54	3.76
2-Hexyl, 4-bromo	6.015	4.26	4.22
2-Cyclohexyl, 4-bromo	5.559[b]	3.75	3.86
2-Bromo	3.043	1.78	1.87
2-Bromo, 4-tert-pentyl	5.264	3.39	3.63
2-Bromo, 4-hexyl	6.015	4.11	4.22
2-Bromo, 4-propyl, 3,5-dimethyl	5.354	3.69	3.70

[a] From Klarmann *et al.* [10].
[b] Ring correction applied for cyclohexyl.

TABLE VII

Vapor Toxicities of Alcohols versus the Tomato Plant and the Red Spider

Alcohol	$^1\chi$	Tomato plant, pC		Red spider, pC	
		Obs[a]	Calc	Obs[a]	Calc
Methanol	1.000	2.60	2.62	2.80	2.75
Ethanol	1.414	2.76	2.90	3.00	3.03
Propanol	1.914	3.33	3.25	3.32	3.37
Isopropanol	1.732	3.18	3.12	3.26	3.24
Butanol	2.414	3.69	3.59	3.77	3.70
Pentanol	2.914	4.05	3.93	4.09	4.04
sec-Butanol	2.270	3.46	3.49	3.62	3.60
Isopentanol	2.770	3.95	3.83	4.09	3.94
tert-Pentanol	2.561	3.51	3.69	3.75	3.80
3-Pentanol	2.808	3.69	3.85	3.81	3.97
sec-Pentanol	2.770	3.77	3.83	3.90	3.94
2-Methyl butanol	2.808	3.77	3.86	3.96	3.97
Isobutanol	2.270	3.57	3.49	3.72	3.60
tert-Butanol	2.000	3.41	3.30	3.28	3.42

[a] From Read [11].

Because the bromine atom is present in each molecule except the first, it is not possible to conclude decisively whether the electronic influence of the bromine is a dominant factor. The superiority of the $^1\chi^v$ correlation, however, suggests that the more definitive consideration of the bromine structure, embodied in δ^v, is reflecting a greater role for this atom than just a space-filling substituent.

V. Vapor Toxicities

The toxicity of alcohol vapors against plants and insects has been measured for a series of compounds, as shown in Table VII [11]. Against the tomato plant, the relationship is

$$pC = 0.682 \, ^1\chi + 1.940, \qquad r = 0.963, \quad s = 0.116, \quad N = 14$$

and against the red spider the relationship is

$$pC = 0.673 \, ^1\chi + 2.076, \qquad r = 0.977, \quad s = 0.090, \quad N = 14$$

TABLE VIII

Relative Sweet Taste of Nitroanilines and χ

R	$^1\chi$	$^1\chi^v$	log RS Obs[a]	log RS Calc
—O—CH$_2$—CH$_2$—CH$_3$	6.647	4.264	1.406	1.426
—I	5.109	4.279	0.675	0.899
—O—CH$_2$—CH$_3$	6.147	3.764	0.885	0.904
—Br	5.109	3.556	0.566	0.398
—Cl	5.109	3.166	0.365	0.127
—O—CH$_3$	5.647	3.176	0.293	0.321
—CH$_3$	5.109	3.064	0.337	0.561
—F	5.109	2.452	−0.591	−0.368
—H	4.716	2.653	−0.538	−0.366

[a] From Blanksma and Hoegen [12]. Data converted to molar basis.

In this case the structural detail engendered by the presence of hydroxyl groups (which give a different contribution to the total value of $^1\chi$ relative to $^1\chi^v$) is best described by $^1\chi$. A case could be made for secondary binding, reinforcing the hydroxyl group interaction at a receptor. The variation in the position of the hydroxyl group is more successfully treated by $^1\chi$, suggesting that a mechanism involving hydrogen bonding may be more likely, since this phenomenon would be less dependent on alcohol class than a covalent bond-forming reaction.

VI. Sweet-Tasting Nitroanilines

A series of substituted nitroanilines have been reported to possess sweet taste potency of up to 4000 times the level of sucrose [12]. Kier has studied these and other sweet-tasting molecules and has predicted that the substituent R on these molecules is participating in the drug–receptor interaction in a supporting role [13]. This was concluded to be primarily a dispersion interaction between the R substituent and a receptor feature [14].

The connectivity indices $^1\chi$ and $^1\chi^v$ have been calculated for these molecules and have been compared with the log of the relative sweetness (relative to sucrose) [15]. The results are shown in Table VIII. The regression equation is

$$\log RS = 0.350 \, {}^1\chi + 0.694 \, {}^1\chi^v - 3.856$$

$$r = 0.953, \qquad s = 0.222, \qquad N = 9$$

The wide diversity of the R groups including halogen atoms would make it quite unlikely that ${}^1\chi$ could adequately reflect the group contribution to a phenomenon such as dispersion interaction. The ${}^1\chi$ values for compounds 2, 4, 5, 7, and 8 are identical, whereas there is nearly a twofold range in activity in this set of compounds.

VII. Summary

We have seen in the examples shown in this chapter that values of ${}^1\chi$, ${}^1\chi^v$, or a combination of both may sufficiently characterize the structure of the series of compounds so as to reflect the influence of structural variation on a biological property. With increased emphasis on atoms other than carbon in substituents being varied, the role of ${}^1\chi^v$ in reflecting their structure becomes dominant.

It is possible in selected cases to calculate abbreviated values of ${}^1\chi$, using only the group as a model structure. This has limitations when structural variation is more complex.

The way is certainly open in these studies or in many others to consider the structural descriptions based on extended terms in the connectivity function. These extended terms express various structural characteristics. With sufficiently large lists of compounds with reliable biological data, this higher level approach is a logical extension of studies using only the first-order terms ${}^1\chi$ and ${}^1\chi^v$.

References

1. F. Norrington, R. Hyde, S. Williams, and R. Wooton, *J. Med. Chem.* **18,** 604 (1975).
2. D. J. Crisp and D. Marr, *Proc. Int. Congr. Surface Activity, 2nd,* 310 (1957).
3. W. J. Murray, L. H. Hall, and L. B. Kier, *J. Pharm. Sci.* **64,** 1978 (1975).
4. C. Hansch and W. Dunn, *J. Pharm. Sci.* **61,** 1 (1972).
5. Y. Ichikawa and T. Yamano, *Biochim. Biophys. Acta* **147,** 518 (1967).
6. L. Kier, L. H. Hall, and W. J. Murray, *J. Med. Chem.* **18,** 1272 (1975).
7. B. R. Baker and M. Kazawa, *J. Med. Chem.* **10,** 302, (1967).
8. H. J. Schaeffer, R. N. Johnson, E. Odin, and C. Hansch, *J. Med. Chem.* **13,** 452 (1970).

9. J. M. Clayton and W. P. Purcell, *J. Med. Chem.* **12,** 1087 (1969).
10. E. Klarmann, V. Shternov, L. Gates, and P. Cox, *J. Am. Chem. Soc.* **55,** 4657 (1933).
11. P. Read, *Ann. Appl. Biol.* **19,** 432 (1932).
12. J. Blanksma and D. Hoegen, *Rec. Trav. Chim.* **65,** 333 (1946).
13. L. Kier, *J. Pharm. Sci.* **61,** 1394 (1972).
14. H. D. Holtje and L. Kier, *J. Pharm. Sci.* **63,** 1722 (1974).
15. L. B. Kier, *Eur. Chemoreception Res. Organ. Symp., Wadenswil, Switzerland, 1975.*

Chapter Eight

MULTIPLE CHI TERMS RELATING TO BIOLOGICAL ACTIVITY

A number of biological activities have been related, in congeneric series, to physical properties in a nonlinear form. The most prominent curvilinear relationship has been found between partition coefficient P and biological properties [1,2]. This property–activity relationship has usually been described as a parabola, the equation employed being

$$\log \text{activity} = a \log P + b(\log P)^2 + c \tag{1}$$

Unfortunately, over half the studies in one review were made on sets of data containing eight or fewer observations, meaning that there were only four or less observations per term in Eq. (1). Such use of small data sets is probably of marginal value for the utilization of these equations in establishing a sound basis for further mechanistic interpretation.

Another characteristic of these studies is the fact that, except for a few studies, all were sets of compounds in which a hydrocarbon portion of the molecule was the structural variable. The failure to find this nonlinear characteristic when heteroatom-containing functional group variation is employed in studies suggests a particular role for the hydrocarbon moieties in drug molecule variation. We know from intuition that these moieties are likely to be involved in dispersion interactions with some part of the receptive tissue and to influence solubility in lipid and aqueous phases.

I. The Parabolic Relationship and Partition Coefficient

A number of models have been offered to explain the parabolic nature of the physical property and biological activity relationship employing the partition coefficient [1]. These may be briefly summarized as follows.

A.   The Kinetic Model

It has been suggested that in whole animals a multicompartment model of alternate lipid and aqueous media would exhibit a parabolic relationship between the logarithm of the concentration, log c, and that of the partition coefficient, log P, in the last compartment of a 20-compartment model.

B.   The Thermodynamic Model

This model may be descriptive of the case in which equilibrium is approached between drug in solution and drug at the action site. Even under near equilibrium conditions, it could be expected that a parabolic relationship between log $1/c$ and log P may be likely [3].

C.   Bulk Tolerance

This concept assumes that an increase in one or more dimensions of a drug leads to a crowding or unefficacious fit at a biologically active site.

D.   Active Site Conformation Distortion

This model assumes a gradual transition from agonist to antagonist as the lipophilic character increases in an homologous series. The increased interaction of the longer hydrocarbon portions could result in alteration of either drug or receptor conformation and hence a reduction in the perfection of the fit.

E.   Others

Other models suggested include enhanced metabolism, micelle formation, limited solubility, enzyme poisoning, and insufficient numbers of molecules at a receptor site. The unifying characteristics of these models are the dependence on molecular size of the parabolic relationship with partition coefficient. A more fundamental description of molecular size can be achieved by utilizing a topological index such as molecular connectivity.

TABLE I

Antimicrobial Activity of Alkyl Quaternary Ammonium Salts and χ

R	$^1\chi$	MIC							
		S. aureus		Cl. welchii		P. aeruginosa		Cl. welchii	
		Obs[a]	Calc	Obs[a]	Calc	Obs[a]	Calc	Obs[a]	Calc
C_8H_{17}	8.306	3.69	3.66	3.00	2.68	1.68	1.48	2.69	2.51
C_9H_{19}	8.806	4.20	4.26	3.02	3.28	2.08	2.14	3.02	3.17
$C_{10}H_{21}$	9.306	4.74	4.76	3.57	3.79	2.41	2.68	3.57	3.72
$C_{11}H_{23}$	9.806	5.06	5.16	4.06	4.22	2.92	3.11	4.06	4.19
$C_{12}H_{25}$	10.306	5.61	5.47	4.61	4.56	3.45	3.42	4.61	4.56
$C_{13}H_{27}$	10.806	5.80	5.67	5.10	4.82	3.85	3.61	5.10	4.84
$C_{14}H_{29}$	11.306	5.82	5.77	5.12	4.99	3.96	3.68	5.12	5.02
$C_{15}H_{31}$	11.806	5.84	5.77	5.14	5.08	3.74	3.64	5.14	5.11
$C_{16}H_{33}$	12.306	5.56	5.66	5.16	5.08	3.30	3.48	5.16	5.11
$C_{17}H_{35}$	12.806	5.18	5.46	4.70	5.00	3.06	3.21	4.70	5.01
$C_{18}H_{37}$	13.306	5.19	5.16	4.71	4.83	2.63	2.82	4.71	4.82
$C_{19}H_{39}$	13.806	4.91	4.76	4.73	4.58	2.52	2.31	4.73	4.54

[a] From Marsh and Leake [6].

II.   The Use of a Quadratic Expression in Chi

The relationship between molecular connectivity and partition coefficient has been established [4]. It would appear reasonable that Eq. (1) could be rephrased with χ under many circumstances, so that

$$\log \text{activity} = a^m\chi + b(^m\chi)^2 + c \tag{2}$$

In all probability the order of the $^m\chi$ term would be 1, although a detailed study of other series members is necessary.

Murray *et al*. [5] have recently tested Eq. (2) as a direct replacement for Eq. (1). A few examples are described.

A.   Chi and Homologous Hydrocarbon Series

A simple test of the possible equivalence of Eqs. (1) and (2) comes from the examination of a series of alkylammonium salts (Table I) [1]. The expression using $^1\chi$ is equally as good as that using $\log P$ to relate to the minimum inhibitory concentration (MIC). The results are not too surprising since the chain variation is due to the addition of $-CH_2-$ units with no branching. The increments of both $\log P$ and $^1\chi$ are 0.500. A tabulation of the statistics for the examples in Table I is shown in Table II.

B.   Chi and Branched Hydrocarbon Series

A more rigorous test of the ability of molecular connectivity to reflect structure is the employment of calculated indices of branched substituents to relate to some property. Correlations were reported using Eq. (2) with a series of alkyl carbamates [2]. The membrane and tissue

TABLE II

Equations and Statistical Data for Studies in Table I

MIC $= a(^1\chi)^2 + b^1\chi + c$

Study	a	b	c	r	s	N
S. aureus	−0.201	4.65	−21.1	0.982	0.141	12
Cl. welchii (1)	−0.170	4.11	−19.71	0.966	0.230	12
P. aeruginosa	−0.233	5.31	−26.51	0.962	0.220	12
Cl. welchii (2)	−0.187	4.51	−22.03	0.980	0.190	12

TABLE III

Intestinal Absorption in vitro for Carbamates and χ

R—O—CO—NH₂

R	$^1\chi$	Serosal transfer log(% abs/cm²/hr)		Tissue bound log(% abs/cm²/hr)		Mucosal loss log(% abs/cm²/hr)	
		Obs[a]	Calc	Obs[a]	Calc	Obs[a]	Calc
Methyl	2.270	0.45	0.420	-0.38	-0.405	0.52	0.543
Ethyl	2.770	0.53	0.609	-0.22	-0.176	0.60	0.691
Propyl	3.270	0.71	0.694	-0.04	-0.030	0.78	0.749
tert-Butyl	3.417	0.65	0.699	0.00	-0.003	0.73	0.748
Isobutyl	3.626	0.75	0.690	0.04	0.024	0.83	0.735
Butyl	3.770	0.82	0.674	0.10	0.033	0.90	0.717
tert-Pentyl	3.917	0.68	0.648	0.04	0.036	0.77	0.690
Pentyl	4.270	0.57	0.550	0.02	0.014	0.68	0.594
tert-Hexyl	4.417	0.09	0.494	-0.21	-0.007	0.26	0.541
Hexyl	4.770	0.35	0.322	-0.05	-0.088	0.50	0.382
Heptyl	5.270	-0.17	-0.010	-0.28	-0.272	0.08	0.079
Octyl	5.770	-0.43	-0.447	-0.57	-0.539	-0.19	-0.314
Benzyl	4.900	0.59	0.246	0.01	-0.127	0.69	0.312

[a] From Pan et al. [7].

behavior of these molecules had previously been linked to partition coefficient values through Eq. (1).

The results, shown in Table III and a tabulation of the statistics in Table IV, reveal that $^1\chi$ characterizes enough of the salient structure in the series to relate as well as log P to the absorption properties of these molecules.

C.   Valence Chi Relationships

Thus far the studies have been described for $^1\chi$ and $(^1\chi)^2$ relationships. A few studies are available that permit the employment of a $^1\chi^v$ term to describe the molecular connectivity.

1.   Ether Toxicity against Mice

A series of 25 ethers have been tested for their LD_{50} against mice [6]. The correlation of log P using Eq. (1) has been reported to give $r = 0.864$, $s = 0.208$.

An inspection of the ethers in Table V makes it clear that the use of $^1\chi$ leads to redundant values in several cases. Thus compounds 3 and 11 have identical values of $^1\chi$ but not pC. Similarly, compounds 7 and 13 would also have identical $^1\chi$ values. In both cases the pC difference is about 0.2. The use of $^1\chi^v$ eliminates all but one redundant pair, 15 and 22, which differ in their pC values by only 0.03. The use of $^1\chi^v$ also makes it possible to calculate the connectivity values for the three vinyl groups in the list and to distinguish them from ethyl groups. Using $^1\chi$ results in identical values for compounds 11, 21, and 25. The predicted values for the pC are shown in Table V from the equation

$$pC = 1.444 \, ^1\chi^v - 0.208(^1\chi^v)^2 + 0.462$$

$$r = 0.908, \qquad s = 0.173, \qquad N = 25$$

TABLE IV

Equations and Statistical Data for Studies in Table III

log % abs = $a(^1\chi)^2 + b^1\chi + c$

Study	a	b	c	r	s	N
Serosal transfer	−0.209	1.429	−1.749	0.915	0.168	13
Tissue bound	−0.165	1.291	−2.484	0.938	0.075	13
Mucosal loss	−0.180	1.206	−1.265	0.912	0.144	13

TABLE V

Toxicities of Ethers against Mice and χ

R^1—O—R^2

			pC	
R^1	R^2	$^1\chi^v$	Obs[a]	Calc
CH_3	CH_3	0.816	1.43	1.68
CH_3	C_2H_5	1.404	1.74	2.08
CH_3	C_3H_7	1.904	2.45	2.42
CH_3	iso—C_3H_7	1.799	2.26	2.35
CH_3	cyclo—C_3H_5	1.960	2.75	2.46
CH_3	C_4H_9	2.405	2.70	2.56
CH_3	iso—C_4H_9	2.260	2.79	2.66
CH_3	sec—C_4H_9	2.337	2.79	2.71
CH_3	tert—C_4H_9	2.112	2.79	2.56
CH_3	C_5H_{11}	2.905	2.88	2.81
C_2H_5	C_2H_5	1.992	2.22	2.48
C_2H_5	C_3H_7	2.492	2.60	2.61
C_2H_5	iso—C_3H_7	2.386	2.60	2.74
C_2H_5	cyclo—C_3H_5	2.548	3.00	2.85
C_2H_5	C_4H_9	2.992	2.82	2.87
C_2H_5	iso—C_4H_9	2.847	2.82	2.73
C_2H_5	sec—C_4H_9	2.924	2.85	2.94
C_2H_5	tert—C_4H_9	2.700	2.92	2.96
C_2H_5	C_5H_{11}	3.492	3.00	3.10
C_2H_5	tert—C_5H_{11}	3.261	3.15	3.19
C_2H_5	$CH_2{=}CH—$	1.640	2.34	2.24
C_3H_7	C_3H_7	2.992	2.79	2.86
C_3H_7	iso—C_3H_7	2.886	2.82	2.75
iso-C_3H_7	iso—C_3H_7	2.781	2.82	3.01
$CH_2{=}CH—$	$CH_2{=}CH—$	1.288	2.33	2.00

[a] From Shonle and Moment [8].

The results are superior to the log P relationship. This is quite encouraging in view of the presence of 11 branched substituents, 3 double-bonded groups, 2 cycles plus a heteroatom in each molecule. It would appear in this case that $^1\chi^v$ is describing structural characteristics that explain 82% of the variation of the pC values in the series.

2.   Mouse Hypnosis by Propargynols

In a second case, the hypnotic dose of propargynols against mice [7], the use of $^1\chi^v$ is obviously necessary. The set of compounds contain

TABLE VI

$$R^1$$
$$|$$
Hypnotic Activity of Propargynols against Mice versus χ HO—C—C≡C—X
$$|$$
$$R^2$$

R^1	R^2	X	$^1\chi^v$	pC	
				Obs[a]	Calc
CH_3	Vinyl	H	1.959	2.41	2.36
CH_3	C_2H_5	H	2.323	2.59	2.68
C_2H_5	Vinyl	H	2.520	2.79	2.82
C_3H_7	Chlorovinyl	H	3.640	2.90	3.03
iso-C_3H_7	Vinyl	H	2.903	2.92	3.00
CH_3	C_2H_5	Cl	2.886	2.94	2.99
CH_3	Chlorovinyl	H	2.579	2.94	2.86
iso-C_3H_7	Chlorovinyl	H	2.523	3.17	3.05
C_2H_5	Chlorovinyl	H	3.140	3.20	3.06

[a] From Kier et al. [9].

double and triple bonds as well as chlorine atoms in half the cases. No adequate differentiation is possible using $^1\chi$. The calculated values are shown in Table VI for the equation

$$pC = 2.535\ ^1\chi^v - 0.381(^1\chi^v)^2 - 1.147, \quad r = 0.917, \quad s = 0.116, \quad n = 9$$

Even with this small number of observations it is satisfying to see that the choice of δ^v for chlorine is adequate to permit a direct comparison with carbon atoms in a series of compounds.

D.   Use of Group Chi Values in Activity Comparisons

We have discussed the possibility of using $^m\chi$ values only for substituent groups in the preceding chapter. The inherent problems are: (1) the relative deviation from whole-molecule $^m\chi$ values when m is greater than 1, and (2) the limited use that substituent $^m\chi$ values have when alternate positioning of substituents in a molecule is part of the study.

To illustrate more graphically the former problem, let us consider a series of cyclohexane derivatives (I). If we calculate the whole-molecule $^1\chi$ for each member of the series, the value will increase by 0.5 when R is greater than ethyl in straight-chain substitution. Similarly, if we consider the $^1\chi$ values only for the substituents R, the $^1\chi$ will also increase by 0.5 for each methylene group. The relative values will be close enough to permit the use of substituent $^1\chi$ values in relating structure to activity.

(I)

This parallel is not found when $^3\chi$ or higher is being considered. In the whole-molecule calculation of $^3\chi$, the ring plays an integral part in contributing the set of δ values making up $^3\chi$. Not until R is C_5 or larger in a straight-chain series do subsequent group $^3\chi$ values parallel whole-molecule $^3\chi$ values. In spite of this, there may be enough information conveyed by calculating group $^m\chi$ values to consider its use in evaluating certain selected structures for SAR studies. One opportunity presents itself to test the feasibility of using group χ values in SAR work.

1. Connectivity Relationships among Barbiturates

Murray *et al.* [5] have examined the relationships of $^1\chi$ values to various activities of barbiturates [8] according to Eq. (2). The results of the prediction of pC values are shown in Table VII from the equation

$$pC = 4.849 \, ^1\chi - 0.328(^1\chi)^2 - 14.26, \quad r = 0.983, \quad s = 0.067, \quad N = 12$$

This expression was derived from calculations of $^1\chi$ using the entire barbiturate molecule. With partition coefficient, the relationship derived from Eq. (1) results in $r = 0.841$, $s = 0.100$.

As an alternative approach, a consideration might be given to the calculation of the group $^1\chi$ values for the partial structure II, where δ for

(II)

the central carbon atom is 4. This δ value is constant through the series; hence group $^1\chi$ values parallel the whole-molecule $^1\chi$ values. Comparison of group $^1\chi$ values with pC values using Eq. (2) gives the expression

$$pC = 2.260 \, ^1\chi - 0.327(^1\chi)^2 - 0.234, \quad r = 0.983, \quad s = 0.067, \quad N = 12$$

The predicted values from this equation are also shown in Table VII. As expected, the partial or group $^1\chi$ relationship is virtually identical to

TABLE VII

Barbiturate Activity and χ Using a Quadratic Expression

R	R	Group $^1\chi$	pC Obs[a]	pC Calc
Ethyl	Ethyl	2.121	3.09	3.09
Ethyl	Isopropyl	2.504	3.30	3.37
Ethyl	Butyl	3.121	3.72	3.63
Ethyl	Isobutyl	2.977	3.63	3.59
Ethyl	sec—Butyl	3.042	3.63	3.61
Ethyl	Isopentyl	3.477	3.75	3.67
Ethyl	Phenyl	4.166	3.46	3.50
Propyl	Phenyl	3.121	3.55	3.63
Propyl	Isopropyl	3.004	3.63	3.60
Propyl	Isopentyl	3.977	3.48	3.58
Propyl	Benzyl	5.139	2.76	2.74
Isopropyl	Benzyl	5.022	2.87	2.86

[a] From Shonle and Moment [8].

the relationship involving the $^1\chi$ for the entire molecule. The abbreviated $^1\chi$ values, considering only the varying R groups, give a very satisfactory partial structure description of the molecule such that in quadratic form the description correlates well with the biological activity.

III.   Alternative to Quadratic Expressions in SAR

It has become a standard procedure to treat apparently nonlinear biological activity data as a quadratic relationship with partition coefficient in a series of compounds. We have very briefly described several mechanistic explanations for the observation of a parabolic relationship linking biological activity and a physical property.

No matter what mechanism is chosen to explain the calculated relationship, it is axiomatic that the independent variable is the structure. This fundamental fact underlies all of the perceived relationships between pairs of dependent/independent variables, whether they are partition coefficient and biological activity or other pairs, such as those discussed in Chapters 4 and 5. In the final analysis, when we can adequately define the salient structures influencing a property, we will obtain a good mathematical correlation with that property.

We have previously described physical properties that are apparently nonlinearly related to a parameter such as the number of carbon atoms. One case in point is the density (specific gravity) of hydrocarbons [9]. We have not attempted to relate the value of density to χ in a quadratic form, although this would be an interesting exercise. The fact is that within the framework of molecular connectivity theory as we now perceive it, $(^1\chi)^2$ has no structural significance. The fact that $(^1\chi)^2$ is useful in the biological studies described in this chapter attests primarily to its purely mathematical role in paralleling some structural description so that a quadratic equation correlates satisfactorily. The same caveat applies to $(\log P)^2$.

TABLE VIII

Extended Connectivity Terms for Ethers Listed in Table V

R^1—O—R^2

R^1	R^2	$^1\chi^v$	$^2\chi^v$	$^3\chi^v$	$^4\chi^v$	$^5\chi^v$
CH_3	CH_3	0.816	0.408	0	0	0
CH_3	C_2H_5	1.404	0.577	0.289	0	0
CH_3	C_3H_7	1.904	0.993	0.408	0.204	0
CH_3	iso-C_3H_7	1.799	0.707	0.471	0	0
CH_3	cyclo-C_3H_5	1.960	0.569	0.333	0	0
CH_3	C_4H_9	2.405	1.346	0.702	0.289	0.144
CH_3	iso-C_4H_9	2.260	1.272	0.500	0.333	0
CH_3	sec-C_4H_9	2.337	1.455	0.977	0.167	0
CH_3	tert-C_4H_9	2.112	2.316	0.612	0	0
CH_3	C_5H_{11}	2.905	1.700	0.702	0.496	0.204
C_2H_5	C_2H_5	1.992	0.781	0.408	0.204	0
C_2H_5	C_3H_7	2.492	1.197	0.553	0.289	0.144
C_2H_5	iso-C_3H_7	2.386	1.260	0.500	0.333	0
C_2H_5	cyclo-C_3H_5	2.548	0.789	0.402	0.236	0
C_2H_5	C_4H_9	2.992	1.551	0.743	0.391	0.204
C_2H_5	iso-C_4H_9	2.847	2.053	0.655	0.354	0.236
C_2H_5	sec-C_4H_9	2.924	1.674	1.026	0.402	0.118
C_2H_5	tert-C_4H_9	2.700	2.545	0.577	0.433	0
C_2H_5	C_5H_{11}	3.492	1.905	1.096	0.598	0.276
C_2H_5	tert-C_5H_{11}	3.261	2.046	1.387	0.493	0.102
C_2H_5	CH_2=CH—	1.640	0.622	0.285	0.118	0
C_3H_7	C_3H_7	2.992	1.612	0.697	0.391	0.204
C_3H_7	iso-C_3H_7	2.886	1.920	0.655	0.354	0.236
iso-C_3H_7	iso-C_3H_7	2.781	2.234	0.544	0.544	0
CH_2=CH—	CH_2=CH—	1.288	0.469	0.236	0.068	0

Our approach to the connectivity description of structure, to reflect the property of density, has been to examine the contribution of various members of an expanded series of $^m\chi$. We have found a good correlation with density based on $1/^1\chi$ and $^3\chi$. This is not a quadratic expression although it cannot be said that $^1\chi$ and $^3\chi$ are completely independent either. They are complexly related terms, which in turn are interrelated to all of the other $^m\chi$ terms describing the connectivity of the molecule. (See Appendix B.) As an example, we have found that the total number of path terms making up all $^m\chi$ terms for an alicyclic hydrocarbon is a constant depending only on the number of carbon atoms.

With the success of using several $^m\chi$ terms to describe structural influence on a property exhibiting nonlinearity, such as density, we are encouraged to examine the ability of this approach to mirror parabolic biological activities. Two examples illustrate this approach.

A.   Ether Toxicity and $^m\chi$ Terms

Earlier in this chapter we demonstrated that a quadratic expression in $^1\chi^v$ gave a fair correlation with the pC for 25 ethers ($r = 0.908$, $s = 0.173$). The $^m\chi$ terms, $^2\chi$ through $^6\chi$ (shown in Table VIII) were compared through multiple regression with $^1\chi^v$. The calculation reveals that $^1\chi^v$ and $^5\chi^v$ give the best correlation from the equation

$$pC = 0.676\ ^1\chi^v - 1.403\ ^5\chi^v + 1.132, \quad r = 0.904, \quad s = 0.177, \quad N = 25$$

The correlation is almost identical to the quadratic relationship.

It is interesting to note that the quality of the correlation with $^1\chi^v$ and $^m\chi^v$ increase and maximize at $^5\chi^v$ and then decline. The $^5\chi^v$ term can be said to embody within it a structural definition that is important in the ultimate measured biological activity.

B.   Barbiturate Activity and $^m\chi$ Terms

The barbiturate activity data shown in Table VII provide another illustration of a nonquadratic equation relating $^1\chi$ and a $^m\chi$ term with biological activity. In this case the $^1\chi$ and $^7\chi$ terms, calculated for the abbreviated portion of the molecule containing the substituent groups, correlated well with the activity. The relation is

$$pC = 0.533\ ^1\chi - 5.800\ ^7\chi_P + 2.006, \quad r = 0.934, \quad s = 0.131, \quad N = 12$$

This relationship is somewhat inferior to the quadratic expression in χ but is nevertheless a good correlation.

IV. Summary

The results in this chapter may be interpreted at two levels. In the first level we have shown that $^1\chi$ and $^1\chi^v$ may be manipulated mathematically to simulate a quadratic dependence on biological activity. Such results mirror the results of work in which a quadratic dependence on the logarithm of partition coefficient has yielded good correlation. In this sense we have shown additional versatility in the application of χ.

It must be emphasized that the parabolic relation to log P is achieved because of the choice of the independent variable in the expression $y = a(\log P)^2 + b \log P + c$. Intrinsically, however, there is no demonstrated basis for a parabolic relation between structure and activity. When suitable descriptors of structure are generated, then a linear dependence of activity on structure descriptors may be observed.

We have found two studies in this chapter that provide an encouraging basis for further work. A high-order extended term has been blended with a first-order term into a multiple linear relationship with biological activity. Such results may open the way to new interpretations.

In the case of ether toxicity, the $^5\chi^v$ term is the best second variable with $^1\chi^v$. When $^5\chi^v$ is nonzero, deviation from linear dependence on $^1\chi^v$ is described by the $^5\chi^v$ value. The $^5\chi^v$ term becomes significant for larger molecules, which contain at least one path with five edges. The $^5\chi^v$ term is zero for short-chain molecules and highly compact or highly branched molecules.

These observations shed some new light on what has been considered simply a quadratic dependence on partitioning. In addition, the way may now be open for the consideration of more complex structures and the identification of structural features that play a significant role in determining activity.

References

1. C. Hansch and J. Clayton, *J. Pharm. Sci.* **62**, 1 (1973).
2. E. J. Lien, *in* "Medicinal Chemistry IV," *Proc. Int. Symp. 4th* (J. Maas, ed.), p. 319, Noordwijkerhout, Netherlands, 1974.
3. T. Higuchi and S. S. Davis, *J. Pharm. Sci.* **59**, 1376 (1970).
4. W. J. Murray, L. H. Hall, and L. B. Kier, *J. Pharm. Sci.* **64**, 1978 (1975).
5. W. J. Murray, L. B. Kier, and L. H. Hall, *J. Med. Chem.* **19**, 573 (1976).
6. D. F. Marsh and C. D. Leake, *Anesthesiology* **11**, 455 (1950).
7. S. Y. Pan, L. Markarian, W. M. McLamore, and A. Bavley, *J. Pharmacol. Exp. Ther.* **109**, 268 (1953).
8. H. A. Shonle and A. Moment, *J. Am. Chem. Soc.* **45**, 243 (1923).
9. L. B. Kier, L. H. Hall, W. J. Murray, and M. Randić, *J. Pharm. Sci.* (in press).

Chapter Nine

USE OF PHYSICAL PROPERTY TERMS WITH CONNECTIVITY FUNCTIONS

We have seen in the last three chapters that structural definitions through molecular connectivity may provide enough information to permit good correlations with biological activities. The electronic structures of variable parts of molecules in a series are not predicted to be significant in the critical steps leading to the dose–response value, or they are partially subsumed into the connectivity description.

Alternatively, there must be cases where the electronic structure of selected parts of the drug molecule must be explicitly considered. In a hypothetical situation, if the acid dissociation of substituted benzoic acid constitutes the rate-limiting process leading to a measured response, then the electronic influence on the carboxyl group by the ring substituent is the salient property to consider in an SAR study. Accordingly we would select a molecular orbital description of the influence, or perhaps an empirical measure of electronic influence such as the Hammett sigma constant.

Few drug–receptor or drug–membrane interactions are likely to depend exclusively on one structural characteristic, be it electronic or topological. Indeed, there are numerous examples of drug action values that have been shown to relate to both a Hammett sigma value and to a physical property, such as partition coefficient [1].

It is apparent then that the expectation of the structure–activity relationship using molecular connectivity may be fullfilled in these cases only by including a descriptor of electronic structure. Several examples can be presented to illustrate this application of molecular connectivity.

I.   Inhibition of *Aspergilus Niger*

A number of substituted benzyl alcohols have been shown to exhibit an inhibitory influence on the mold *Aspergilus niger* [2]. The potency of these molecules, expressed as log $1/C$ (concentration) is shown in Table I. Considering the $^1\chi^v$ values for each molecule, a fair correlation is found between these and log $1/C$ ($r = 0.890$). The electronic influence of the substituents on the ring or on the carbinol is an additional factor, however, which must be considered in establishing an SAR for this system. A multiple regression including $^1\chi^v$ and the Hammett sigma values for the ring substituents leads to the following relationship and statistics:

$$\log 1/C = 0.910 \ ^1\chi^v + 0.595\sigma - 0.930$$

$$r = 0.937, \qquad s = 0.286, \qquad N = 19$$

TABLE I

Inhibition of Aspergilus Niger by Substituted Benzyl Alcohols versus χ and Hammett Sigma Terms

			log $1/C$	
R	σ	$^4\chi_P^v$	Obs	Calc
H	0	0.574	1.51	1.51
4-Cl	0.23	0.718	2.07	1.91
2,4-Cl$_2$	0.45	1.153	3.07	2.88
3,4-Cl$_2$	0.60	0.980	3.07	2.61
2,4,5-Cl$_3$	0.82	1.365	3.32	3.49
3,4,5-Cl$_3$	0.97	1.400	3.63	3.64
2-Br	0.20	1.024	2.15	2.50
4-Br	0.23	0.851	2.27	2.17
4-I	0.28	1.107	2.75	2.70
4-Me	−0.17	0.684	1.79	1.64
2,4-Me$_2$	−0.31	1.008	2.14	2.21
3,5-Me$_2$, 4-Cl	0.09	1.238	3.05	2.87
3,5-Me$_2$, 4-I	0.14	1.575	3.42	3.56
2-NO$_2$	0.76	0.786	2.49	2.31
4-NO$_2$	0.78	0.697	2.00	2.14
4-CN	0.63	0.676	1.67	2.03
2-OH	−0.09	0.623	1.39	1.54
3-OH	0.12	0.635	1.39	1.67
4-OH	−0.37	0.586	1.39	1.32

It should not become a standard procedure in this work to presume that $^1\chi$ is always necessary. We have found in this study that the $^4\chi_P^v$ and sigma terms give a superior correlation with activity:

$$\log 1/C = 2.021\ ^4\chi_P^v\ + 0.504\,\sigma + 0.323$$

$$r = 0.959, \qquad s = 0.225, \qquad N = 19$$

The predicted values for log $1/C$ from this equation are shown in Table I. The equation reveals a meaningful contribution from both the connectivity and the electronic terms.

A possible explanation for the dual influence of χ and σ on the measured activity may be the necessity in the rate-limiting mechanism for a labile proton on the carbinol. This may be coupled with a secondary dispersion interaction at the aromatic ring and its periphery. Thus the χ term reflects the connectivity and therefore the potential for dispersion interaction at the aryl portion of the molecule.

Certainly the relationships developed for this system is open to other interpretations. Instead of a simultaneous interaction of two parts of a molecule in a rate-limiting step, the indices χ and σ may be characterizing the same structural feature in somewhat different ways. This structural feature is thus more completely characterized by χ and σ in concert, so that a significant correlation may be obtained. It is encouraging to see that the connectivity index in concert with the Hammett substituent value can reveal such a good relationship with the biological activity.

II. The Toxicity of Diethyl Phenyl Phosphates

A series of ring-substituted diethyl phenyl phosphates have been shown to be toxic to the housefly [3]. The series of molecules demonstrates a combined dependence of the activity on the electronic influence on the ring, and the connectivity description of the substituent. The equation relating the toxicity pC and the structural variables is

$$pC = 2.512\sigma + 0.382\ ^1\chi^v - 1.350, \qquad r = 0.975, \quad s = 0.295, \quad N = 13$$

The experimental and predicted values from this equation are shown in Table II.

We may speculate here that the electronic term is describing the substituent influence on the reactivity of the phosphate residue, while the $^1\chi^v$ term is describing a connectivity characteristic influencing the possible dispersion influence of the molecule in the region of the

TABLE II

Toxicity of Substituted Phenyl Diethyl Phosphates versus Hammett Sigma and χ

| Phenyl substituent | σ | $\chi^{v \, a}$ | pC | |
			Obs	Calc
4-NO$_2$	1.24	2.358	2.40	2.66
4-SO$_2$CH$_3$	0.92	3.046	2.10	2.12
4-CN	0.88	2.295	1.85	1.74
3-NO$_2$	0.71	2.358	1.48	1.33
3-SF$_5$	0.68	1.495	1.38	0.94
4-Cl	0.23	2.423	0.04	0.15
3-*tert*-C$_4$H$_9$	−0.10	3.571	−0.24	−0.24
H	0	1.911	−0.82	−0.62
4-COOH	0.35	2.499	0.25	0.48
4-*tert*-C$_4$H$_9$	−0.20	3.571	−0.24	−0.49
3-O-CH$_3$	0.12	2.433	−0.25	−0.12
4-O-CH$_3$	−0.11	2.433	−0.25	−0.69
4-CH$_3$	−0.17	2.321	−1.32	−0.89

^{a} Calculated for phenyl ring and substituent only.

aromatic ring. Apparently both structural features must be optimized in order to obtain the measured activity.

III.   Inhibition of *A. Niger* by Substituted Phenols

The mold *A. niger* is inhibited by variously substituted phenols as shown in the study by Shirk *et al.* [4]. Correlation of the potency pC, expressed as a relative concentration, with $^1\chi^v$ is unsatisfactory. A parabolic relation is indicated. Further, an electronic effect is suggested by the significance of the Hammett sigma term in a multiple regression with $^1\chi^v$ and its square:

$$pC = -2.94 + 3.18\,^1\chi^v - 0.343(^1\chi^v)^2 + 1.08\sigma$$

$$r = 0.958, \qquad s = 0.21, \qquad N = 18$$

Observed and calculated results appear in Table III. Statistical analysis reveals that all three variables are approximately equal in significance, with $^1\chi^v$ slightly more so than the other two.

The significance of the sigma term may suggest the fundamental importance of the electronic effect on events outside the cell, such as

TABLE III

Inhibition of Aspergilus Niger by Substituted Phenols versus Hammett Sigma and χ

Substituents	σ	χ^v	log $1/C$	
			Obs	Calc
H	0	2.134	2.35	2.28
4-Cl	0.23	2.647	3.35	3.32
2-Me	−0.13	2.551	2.70	2.80
2-Me, 4-Cl	0.10	3.064	3.70	3.69
3-Me	−0.07	2.545	2.68	2.85
3-Me, 4-Cl	0.16	3.064	3.70	3.75
2,6-Me$_2$	−0.28	2.968	3.35	3.17
2,6-Me$_2$, 4-Cl	−0.03	3.481	4.40	3.94
3,5-Me$_2$	−0.14	2.956	3.26	3.30
3,5-Me$_2$, 4-Cl	0.09	3.481	4.22	4.07
2-iso-C$_3$H$_7$	−0.23	3.494	3.35	3.73
2-iso-C$_3$H$_7$, 4-Cl	0	4.007	4.30	4.29
3-Me, 6-C$_4$H$_9$	−0.59	4.212	3.70	3.73
3-Me, 4-Cl, 6-*tert*-C$_4$H$_9$	−0.36	4.730	4.30	4.04
2-cyclo-C$_6$H$_{11}$	−0.23	4.656	4.00	4.18
2-cyclo-C$_6$H$_{11}$, 4-Cl	0	4.932	4.40	4.40
2-Phenyl	0	3.859	4.10	4.22
2-Phenyl, 4-Cl	0.23	4.392	4.52	4.65

dissociation of the acidic phenol proton. Certainly such electronic effects influence physiologically important processes such as passage through cell walls.

On the other hand, the combination of a connectivity term, reflecting a whole-molecule influence, with the sigma term, which relates to the substituent group, suggests that both may be important at the active site. The $^1\chi^v$ term may relate to the interactions of the substituted ring over a broad area, whereas σ may relate to a more local electronic effect.

IV.   Chlorosis in *Lemna Minor* by Phenols

Blackman and his co-workers studied the phenol-induced chlorosis in *Lemna minor* [5]. Their relative potencies are expressed as equipotent concentration pC. In their analysis an attempt was made to relate pC to the phenol pK_a and solubility. Limited correlations were found.

Correlation of $^1\chi^v$ with pC results in $r = 0.917$. The addition of either the Hammett sigma term or the pK_a value reduces the standard error about 40%:

$$pC = -0.55 + 1.116\,{}^1\chi^v + 0.603\sigma, \qquad r = 0.958, \quad s = 0.34, \quad N = 25$$

$$pC = 2.122 + 1.114\,{}^1\chi^v - 0.273(pK_a)$$

$$r = 0.965, \qquad s = 0.3\mathrm{i}, \qquad N = 25$$

As one might expect, σ and pK_a are highly correlated, $r = 0.975$. Since pK_a is an experimental quantity requiring synthesis of the actual

TABLE IV

Phenol-Induced Chlorosis in Lemma Minor versus Hammett Sigma and χ

Phenol substituent	χ^v	σ	pC Obs	pC Calc
H	2.134	0	1.80	1.83
4-Cl	2.647	0.23	2.70	2.54
2,4-Cl$_2$	3.165	0.60	3.40	3.34
2,4,6-Cl$_3$	3.684	0.97	4.50	4.14
2,4,5-Cl$_3$	3.684	1.27	5.10	4.32
2,3,4,6-Cl$_4$	4.209	1.96	5.60	5.33
Cl$_5$	4.734	2.32	6.20	6.13
2-CH$_3$, 4-Cl	3.064	0.10	3.20	2.93
2-CH$_3$, 6-Cl	3.070	0.55	2.40	3.20
3-CH$_3$, 4-Cl	3.064	0.16	3.20	2.96
4-CH$_3$, 2,6-Cl$_2$	3.582	1.18	3.40	4.15
3-CH$_3$, 2,4,6-Cl$_3$	4.107	1.51	4.70	4.94
3-CH$_3$, 2,4,5,6-Cl$_4$	4.632	1.88	5.40	5.75
2-CH$_3$	2.551	-0.13	2.20	2.21
2,6-(CH$_3$)$_2$	2.968	-0.26	2.40	2.60
2,4-(CH$_3$)$_2$	2.962	-0.30	2.60	2.57
2,5-(CH$_3$)$_2$	2.968	-0.20	2.60	2.63
3,5-(CH$_3$)$_2$	2.956	-0.14	2.70	2.66
2,4,6-(CH$_3$)$_3$	3.379	-0.43	2.80	2.95
2,3,5-(CH$_3$)$_2$	3.379	-0.27	3.00	3.05
3-CH$_3$, 5-C$_2$H$_5$	3.516	-0.14	3.10	3.28
3,5-(CH$_3$)$_2$, 4-Cl	3.480	0.09	3.60	3.38
2,5-(CH$_3$)$_2$, 4-Cl	3.480	0.03	3.60	3.35
2,6-(CH$_3$)$_2$, 4-Cl	3.480	-0.03	3.40	3.31
3-CH$_3$, 5-C$_2$H$_5$, 4-Cl	4.041	0.09	4.00	4.01

molecule for its measurement, the process of drug design may be better served by the use of two theoretically derived quantities such as χ and σ. Since the standard errors for the two above equations are not significantly different, the first one is therefore preferable. Observed and calculated values based on the first equation are shown in Table IV.

V. Inhibition of *T. Mentagrophytes*

A series of substituted phenyl propyl ethers have been tested for their antifungal activity against *T. mentagrophytes* (Table V) [6].

In a study of these data, Hansch and Lien [7] found a correlation with log *P* (partition coefficient) and the Hammett sigma value ($r = 0.911$, $s = 0.216$). He omitted two observations from his regression analysis. We have examined these data using terms of χ. We found that σ made no significant contribution to a relationship. A three-term equation, considering all 28 molecules, gave

$$\log 1/C = 2.45 \, {}^1\chi - 3.29 \, {}^3\chi_P + 2.70 \, {}^4\chi_{PC}^v - 1.32$$

$$r = 0.957, \qquad s = 0.149, \qquad N = 28$$

Adding either a ${}^4\chi_P^v$ or a ${}^3\chi_C^v$ term improves the relationship:

$$\log 1/C = 3.73 \, {}^1\chi - 4.34 \, {}^3\chi_C^v - 0.61 \, {}^4\chi_P^v + 3.05 \, {}^4\chi_{PC}^v - 4.69$$

$$r = 0.970, \qquad s = 0.128, \qquad N = 28$$

This study indicates the additional power of the molecular connectivity method. There are several topological descriptors available to describe the relation between structure and properties. In this case one or two additional terms make the difference between a good correlation and an excellent correlation, allowing the inclusion of what otherwise appeared to be discordant observations. One of the limitations of the PAR methods is thus overcome in a straightforward manner by molecular connectivity.

VI. Summary

These examples illustrate the compatability of the molecular connectivity indices with the electronic indices such as the Hammett sigma value in correlating with biological activities.

This capability permits the use of molecular connectivity in SAR studies in which nontopological influences are contributing to activity.

TABLE V

Antifungal Activity of Phenyl Propyl Ethers versus Functions of χ

R	X	Y	log 1/C Obs	log 1/C Calc
2-CH$_3$	OH	OH	2.26	2.07
2-CH$_3$	OH	H	2.46	2.63
2-CH$_3$	H	OH	2.79	2.64
2-Cl	OH	OH	2.31	2.21
2-Cl	OH	H	2.84	2.78
4-Cl	OH	OH	2.31	2.42
4-Cl	OH	H	2.81	2.99
4-Cl	H	OH	3.07	3.00
2,6-Cl$_2$	OH	OH	2.37	2.39
2,6-Cl$_2$	OH	H	3.04	2.95
2,4-Cl$_2$	OH	H	3.35	3.27
2,4-Cl$_2$	OH	OH	2.61	2.71
2-CH$_3$, 4-Cl	OH	H	3.30	3.15
2-CH$_3$, 4-Cl	OH	OH	2.33	2.59
3-CH$_3$, 4-Cl	OH	OH	2.90	2.80
3-CH$_3$, 4-Cl	OH	H	3.30	3.36
2-CH$_3$, 6-Cl	OH	OH	2.33	2.27
2-CH$_3$, 6-Cl	OH	H	2.70	2.83
2-CH$_3$, 6-Cl	H	OH	2.78	2.85
2,6-(CH$_3$)$_2$, 4-Cl	OH	OH	2.76	2.89
2,6-(CH$_3$)$_2$, 4-Cl	OH	H	3.51	3.45
2,6-(CH$_3$)$_2$, 4-Cl	H	OH	3.51	3.46
3,5-(CH$_3$)$_2$, 4-Cl	OH	OH	3.24	3.22
2,6-(CH$_3$)$_2$, 4-CH$_3$	OH	OH	3.10	3.03
3,5-(CH$_3$)$_2$, 4-Cl	OH	H	3.68	3.79
2,6-(CH$_3$)$_2$, 4-CH$_3$	OH	H	3.47	3.59
3,5-(CH$_3$)$_2$, 4-Cl	H	OH	3.93	3.80
2,6-Cl$_2$, 4-CH$_3$	H	OH	3.67	3.60

The assumption is made that connectivity and electronic structure are linearly independent. In the next chapter we demonstrate some dependence between these two in the case of hydrocarbons. In the case of heteroatom molecules, as usually considered in medicinal chemistry, the application of simple connectivity to electronic properties is insufficient for a meaningful description. Thus χ and an electronic term, in concert, is a useful pair of structural descriptors for drug SAR studies.

References

1. W. van Valkenburg (ed.), "Biological Correlation—The Hansch Approach," *Advan. Chem. Ser. 114*. Amer. Chem. Soc., Washington, 1972.
2. D. V. Carter, P. T. Charlton, A. H. Fenton, J. R. Housey, and B. Lessel, *J. Pharm. Pharmacol.* **10,** *Suppl.,* 149T (1958).
3. R. L. Metcalf and T. R. Fukuto, *J. Econ. Entomol.* **55,** 340 (1962).
4. H. G. Shirk, R. R. Corey, and P. L. Poelma, *Arch. Biochem. Biophys.* **32,** 392 (1951).
5. G. E. Blackman, M. H. Parke, and G. Garton, *Arch. Biochem. Biophys.* **54,** 45, 55 (1955).
6. F. M. Berger, C. V. Hubbard, and B. J. Ludwig, *Appl. Microbiol.* **1,** 146 (1953).
7. C. Hansch and E. J. Lien, *J. Med. Chem.* **14,** 653 (1971).

Chapter Ten

REFLECTIONS ON THE NATURE AND FUTURE OF CONNECTIVITY

The preceding chapters have revealed the relationships between functions of χ and physical or biological properties. The question is, what is chi and why does it perform so well? In this last chapter, we will summarize some of our observations and impressions and will reflect on the meaning of these. We will also describe some incomplete work in this area in the hope that others will be stimulated to find new answers.

I. The Formulation of χ

χ is a weighted count of subgraphs. The graph is decomposed into subgraphs of specified numbers of edges. The subgraphs are numerically described according to the pattern of adjacencies in the graph. The χ for the molecule is a sum or count of subgraph values. These in turn are derived from a weighting based on adjacency patterns. The count of subgraphs and the numerical value of the weighting term depend on molecular structure in different ways. For example, in heptane isomers, the count of one-bond subgraphs is constant while the weighting of each varies. On the other hand, the simple count of two-bond subgraphs varies between 5 and 9.

A formal means of incorporating differences between heteroatoms has been developed in the formulation of δ^v values for the calculation of $^m\chi^v$. The formulation of $^m\chi$ terms does not provide a means to make these distinctions. It is not surprising then to find significant improvement in the correlations with certain properties when $^m\chi$ and $^m\chi^v$ comingle in a multiple regression.

At the level of $^1\chi$, the influence of adjacent atoms only is felt on a particular atom. This occurs through the values of $(\delta_i\delta_j)^{-1/2}$ centering on atom i. Beyond atom j, bonded to atom i, there is no influence expressed in $^1\chi$. This situation is altered in higher order subgraph terms. As a consequence, when a physical property derives its value from these influences, $^m\chi$ functions are necessary.

At the heart of the computation of χ is the atom product $\delta_i\delta_j$. Randić, in formulating $(\delta_i\delta_j)^{-1/2}$ to satisfy inequalities, settled for an expedient. Is a more fundamental rationale for the product, the reciprocal, and the square root extant? The product of deltas is the first step in obtaining a geometric mean. The sum of deltas in a bond would be the first step in deriving an arithmetic mean. Geometric means are clearly more nearly unique numbers than are arithmetic means. The actual value of c_{ij} for each bond is the reciprocal of the geometric mean. The question then arises, should the computation of $^2\chi$ arise from a sum of $(\delta_i\delta_j\delta_k)^{-1/3}$ values rather than the reciprocal square root of the product? This warrants examination.

At this time, we can conclude on an empirical basis that $^m\chi$ encodes certain counts and weights of structural features, which give rise to a unique set of numbers, behaving in such a manner as to reflect the numerical value of many physical properties.

II.   Interpretation of χ Terms

Much of the "SAR" work done in biological studies has dealt with physical properties such as partition coefficient and molar refraction. These physical properties have been found to correlate, in many cases, with biological activities in derived multiple regression equations. As a consequence of the observation of a significant correlation, investigators have expressed a confidence that they have a greater insight into the molecular structure imparting the activity. They may also have a sense of understanding of the mechanism or physicochemical events underlying the biological phenomenon.

This confidence is largely inflated. A correlation of a biological activity with partition coefficient tells us little of the structure of the molecule. The partition coefficient is a function of the structure of the molecule and the consequences of that structure upon interaction in two solvents. Furthermore, the interpretation of underlying mechanisms must deal with the two-solvent partitioning behavior since this is the physical property chosen.

Liberties are taken with these constraints. Investigators have presumed that partition coefficient is an isomorph of some other physicochemical event such as dispersion interaction or "hydrophobic bonding." Out of all of this has developed an intuition about the nature of partition coefficient and what structural characteristics govern it. We reiterate the point made earlier, that these physicochemical properties are not structural descriptors but are themselves complex resultants of structure. They are easily and reproductively measured, and hence are readily available to the investigator.

We have made the point in many ways in earlier chapters that the connectivity indices are a numerical description of structure. They encode within them information about atoms, bonds, and their topological assemblage or connectivity.

The difficulty encountered in considering molecular connectivity is that there is not yet a developed intuition as to what familiar structural characteristics are described by $^1\chi$, $^2\chi^v$, $^5\chi_P^v$, etc. There is no doubt that each term in a series of χ carries information. We have shown in Chapter 5 what value $^1\chi$ and $^3\chi_P$ contribute in correlating with hydrocarbon density.

It is not possible at this time to describe with familiar words what these terms mean. Perhaps a whole new set of words representing structural nuances will have to evolve as pictorial descriptions of each term are developed. Words like branchedness may apply to $^1\chi$ but the language may lack words and our minds may lack images to describe higher term characterizations.

The absence of a clear mental picture of what each term is describing should in no way preclude its use. Indeed we have been using the partition coefficient generously, and probably understand as little about it as connectivity.

At this time we can offer some very preliminary insight into what terms of χ mean. In our recent paper on density, in which we introduced the concept of extended connectivity, we observed that the regression equations for hydrocarbons, alcohols, and acids all had nearly identical constant terms [1]. These equations are

$$\text{Hydrocarbons:} \quad D_4^{20} = 0.7348 + 0.0030\,^3\chi_P - 0.2927/^1\chi$$
$$\text{Alcohols:} \quad D_4^{20} = 0.7933 + 0.0183\,^3\chi_P - 0.0043/^1\chi$$
$$\text{Acids:} \quad D_4^{20} = 0.7546 + 0.0252\,^3\chi_P - 0.4358/^1\chi$$

Furthermore, this constant is very nearly the value of the phase volume (0.7402), which is the fraction of a space occupied by spheres of equal radius when packed most efficiently.

The equations can be analyzed by considering the limiting values of $^1\chi$ and $^3\chi_P$. If $^1\chi$ is infinite and $^3\chi_P$ is zero, both terms drop out and the densities equal the equation constants, or nearly the phase volume.

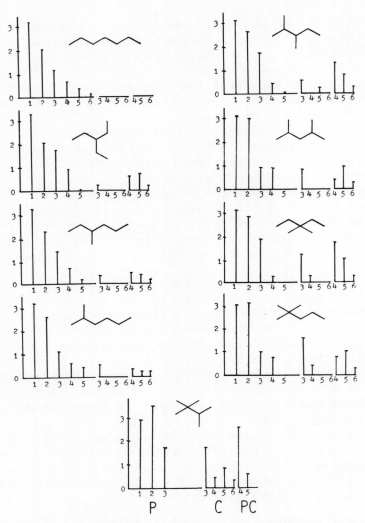

Fig. 1. Bar graph spectra of $^m\chi_t$ values of heptane graphs for each type: P, path; C, cluster; PC, path/cluster. The m value is the horizontal axis in each case and the magnitue for $^m\chi_t$ is the vertical axis.

What kind of a structural characteristic is described by $\lim {}^1\chi \to \infty$ and $\lim {}^3\chi_P \to 0$? If we interject the condition that $\lim {}^2\chi \to \infty$ then the structure described by these connectivity terms is a sphere, an infinite number of connections to a central point ($K_{1,\infty}$ star). The density would accordingly be the phase volume 0.7402 for each of the three types of molecules.

From this analysis, we can at least draw some conclusions about the structural characteristics described by $^1\chi$ and $^2\chi$ when they are infinite. This does not tell us very much about the structure when they are finite numbers.

Another approach to the analysis of structural meaning of $^m\chi$ terms is to compare graphs and values of $^m\chi$ terms to find any discernible trends in structure. One type of listing was developed in Chapter 3, Tables VII and VIII. In another form of analysis we consider the spectrum of $^m\chi_t$ values. Such spectra are shown for heptane isomers in Fig. 1.

We will attempt to provide some generalizations from graphs of alkanes or tree graphs. Perhaps the most striking feature of the set of spectra is the absence of higher order path terms for highly branched structures. This arises because the longest paths in highly branched graphs are shorter. We have pointed out earlier that degree of branching is associated with decreasing $^1\chi$ value. The highest order $^m\chi_P$ term corresponds to an m value,

$$m = n - 1 - b$$

where m is the number of edges, n the number of vertices, and b the number of vertices removed from the straight-chain (unbranched) graph and used in the side-chain branches.

Further, we note that a smooth falloff in the spectrum is associated with graphs with long-chain portions. Graphs with only short chains possess a humped spectrum that is truncated on the high-order side.

Graphs with isolated, remote, or nonadjacent branch points are characterized by high $^2\chi$ and low $^3\chi_P$ values. Interior branching (not on penultimate vertices) creates high $^3\chi_P$ values. Graphs that possess $\delta = 4$ have high $^2\chi$ values because each such branch point gives rise to six two-edge subgraphs. It is also observed that when $^2\chi > {^1\chi}$, three terminal groups ($\delta = 1$) are present. Such a situation indicates the presence of a *tert*-butyl group. For the case in which the molecule is methyl branched at both ends, the $^m\chi$ values for $m = m_{max} - 2$ and $m = m_{max} - 3$ are equal.

In time such analyses of cyclic and other more complex graphs will unfold other relationships. These primitive insights could be developed to the point at which such qualitative descriptions of molecules as "compact," "spherical," or "extended" may be given quantitative expression. Indeed, design of drug molecules may be enhanced as a result of shape interpretations of predicted $^m\chi_t$ values.

III. Comparison of Shape Characteristics

The occasion may arise where connectivity characteristics must be compared between molecules of differing numbers of atoms. It is not

TABLE I

Range of χ/n Values for Structural Types

Structural type	χ/6	χ/10
Cycloalkanes	0.500	0.500
n-Alkanes	0.486	0.491
Methyl cycloalkanes	0.482	0.489
3-Methyl alkanes	0.468	0.481
2-Methyl alkanes	0.462	0.477
1,1-Dimethyl cycloalkanes	0.451	0.471

readily apparent from a list of χ values ranging from 2 to 5 which molecules possess similar connectivity or shape. Accordingly, we may consider the normalization of the χ values to the number of atoms. The $^m\chi/n$ values lend themselves to the search for commonality. Thus a close range of $^1\chi/n$ values may be found for normal alkanes, 2-methyl-alkanes, and so forth. Alkanes from 6 to 10 carbons are analyzed by their $^1\chi/n$ values in Table I.

This normalization should have utility in the classification of apparently dissimilar molecules into comparable structural classes. One example arose in our study of density. Our initial investigation centered on the relationship between hydrocarbons with close $^1\chi/n$ values and their density [1]. We found a nearly perfect linear relationship between $^1\chi/n$ for structural types and the value of their specific gravity.

A second example of the utility of $^1\chi/n$ in relating structures of different numbers of atoms can be found in a consideration of odor-producing molecules. Amoore *et al.* [2] have proposed that molecules possess characteristic primary odors because of space-filling or topological features. This theory lends itself to analysis using $^1\chi/n$. Table II lists several structurally dissimilar molecules with the odor of almonds. Obviously the χ values do not reveal any details of the structure. The $^1\chi/n$ values, however, show a narrow range. This indicates certain common structural features that may be sufficient to impart a common odor.

IV.   Challenging Problems

From our studies, a number of problems have surfaced that continue to challenge our understanding of χ. Work is currently in progress on these. We lay them out for the reader in the expectation that contributions from other investigators may result in solutions.

A. Cis–Trans Isomerism

The graph of a molecule is a one-dimensional picture of the adjacency or connectivity of the atoms. Direction of adjacency is not inherent in the graph. As a consequence, the graphs of *cis*-butene and *trans*-butene are isomorphic (see Chapter 2). The resulting values of $^m\chi_t$ are therefore identical. It is well known that the physical properties are different. As presently formulated, molecular connectivity inadequately describes this characteristic.

We cannot say, at this time, whether the cis–trans isomerism is a topological characteristic, amenable to a solution within the framework we have developed here. It may be that the notions of direction or other spatial relations must be introduced into these alkene graphs as a minimum extension. Progress will be welcomed on this problem.

TABLE II

χ/n Values for Molecules with an Odor of Almonds

Molecular structure	$^1\chi$	$^1\chi/n$
◯– CHO	3.932	0.492
◯– NO₂	4.305	0.478
◯– CN	3.932	0.492
CH₃–◯O◯–CHO	3.826	0.478
◯– CHO	3.932	0.492
◇– CHO	3.308	0.473
◯– CHO	3.808	0.476
▷–CHO	3.843	0.480
◯S– NO₂	3.805	0.476
◯=O	4.394	0.488

B.   Steric Phenomena

Steric interactions or nonbonded effects between atoms play a significant role in the properties of molecules. These three-dimensional characteristics are not explicitly considered in the graphs we have described. There is, however, an implication of steric interactions in the profile of higher order connectivity terms. Thus graphs of molecules with adjacent high-valued delta terms, as in 2,2,3-trimethylheptane, will produce a characteristic spectrum of expanded χ terms. With appropriate χ terms in a multiple regression versus a property, a significant correlation indicates that this implied steric feature has been at least partly incorporated into the connectivity description (see Chapter IV, Section I).

It would be desirable to describe in connectivity terms the steric interaction of groups on molecules that influence physical or biological properties. One approach may be the calculation of partial χ values for groups involved in steric interactions. Empirical treatment of steric effects is incorporated into the Taft steric term [3]. We have found that group χ values correlate with many of the Taft steric values. It would be more desirable to reflect directly the steric influence on a property with a connectivity description.

C.   Conformation

As a further extension of our goal of characterizing steric phenomena with terms of χ, it would be desirable to treat conformational structure. It would seem that a major change in the connectivity formulation must be made to deal with this three-dimensional phenomenon.

D.   Halogen Delta Values

At this time we are utilizing empirical δ^v values for the halogens. These have been calibrated to molar refraction values using δ values of one in a two-term expression. A question arises as to what the fundamental significance of these values may be. Can δ^v values be developed that are nonempirical, within the context of molecular connectivity formalism?

E.   Other Atoms

Other atoms such as phosphorous, sulfur, boron, and metals in covalent molecules are of interest to the biological scientist. The derivation and demonstration of the appropriate delta values for these atoms has not yet been completed.

V. Future Work

There are several areas related to molecular connectivity that deserve further examination. Some of these considerations include alternate formulations of χ, exploration of the dependence of properties on functions of χ, biological applications on systems more complex than those discussed here, and optimum drug molecule design based on connectivity predictions. We examine some of these areas here.

A. The Formulation of χ and the Connectivity Function

It is generally held that the properties of molecules may be described or represented in terms of the various characteristics of the molecular structure. Several investigators have attempted to calculate properties explicitly in terms of the topology of the molecule. As described in Chapters 1 and 2, even the use of bond terms to calculate properties is a topological method. The number of various types of bonds may be determined, prescribed values assigned to each type, and a total value accumulated, all in accordance with the molecular topology. However, such methods often yield only approximate values, sometimes deviating significantly from experimental values, in biology or physical chemistry.

In order to provide better calculation some authors have developed sophisticated schemes for considering larger molecular fragments, variously defined and added into multivariable equations. These are topological methods although some authors have not formally so described their work (Wiener, Platt, Altenburg, Zwolinski). Tatevskii, Hosoya, Herndon, and Smolenski have attempted to develop formal topological schemes primarily for thermochemical data.

All these methods are based on the simple count of structural features, including subgraphs in some cases (Smolenski). We have made a preliminary evaluation of the effect of both the type of subgraphs counted and the type of weighting factor used. For alkanes we have run correlations with several methods for boiling point, heat of vaporization, heat of formation (liquid), and molar refraction. In each case the results were compared to those obtained with the connectivity function as reported in earlier chapters. In every case the connectivity function yielded superior results to methods based on the simple (unweighted) count of subgraphs for one or both of two reasons: (1) a better standard error was obtained regardless of the number of terms used, (2) fewer variables were required to achieve comparable or near comparable results.

The connectivity method also appears superior in the case of heteroatomic molecules. It is not clear at this time, however, how a method based only on the simple count of subgraphs can be extended without greatly increasing the number of variables. Interesting and perhaps fruitful work lies ahead in this area.

One practical advantage of the connectivity method for heteroatoms is the negligible increase in the number of parameters. The heteroatom is introduced simply by modifying the value of the atomic connectivity δ.

We have assumed a linear dependence on each $^m\chi_t$ term in formulating the connectivity function. This is perhaps the simplest approach. However, we have indicated that such properties as density and boiling point are not linear functions of $^1\chi$, even for structurally similar sets in an homologous series. It appears that an examination of the topological dependence of various properties may be a fruitful area. Properties characteristic of isolated molecules may, in fact, require a different functional form than properties characteristic of interaction between molecules.

B. A Problem of High Structural Complexity

Some of the biological studies discussed in this volume contain relatively simple molecules, containing perhaps only two or three significant structural features. However, many of the challenging problems facing the medicinal chemist deal with systems of great structural complexity. We have studied some such systems and discussed them in the four preceding chapters. Here we present an even more complex case.

Baker and Corey have studied the inhibitory action of benzamidine derivatives on guinea pig complement [4]. The set of 108 compounds studied represents a large number of observations on structures containing one to three substituted benzene rings, various lengths and types of ring-bridging chains, and six different heteroatoms in a great variety of functional groups and branching patterns.

We have found a correlation of the inhibitory activity with connectivity. Both $^1\chi$ and $^1\chi^v$ give essentially the same result, $r = 0.900$ and $s = 0.31$:

$$\log 1/C = 0.256 \, ^1\chi^v + 1.695, \qquad r = 0.8997, \quad s = 0.314, \quad N = 108$$

Such a result is most encouraging for this challenging a problem and is much superior to correlation with any single physical property such as partition coefficient or molar refraction, $r = 0.21$ and $r = 0.60$, respectively.

The addition of the $^3\chi_C$ term improves the correlation:

$$\log 1/C = 0.231 \, ^1\chi^v + 15.023 \, ^3\chi_C^v + 1.822$$
$$r = 0.922, \qquad s = 0.28, \qquad N = 108$$

TABLE III

Selected Observed Values for Benzamidine Inhibition of Guinea Pig Complement

Compound	Activity log $1/C^a$
1. —H	2.39
2. 3,5-(OCH$_3$)$_2$	2.21
3. 3-OH	2.41
4. 3-CF$_3$	2.44
5. 3-O(CH$_2$)OC$_6$H$_5$	3.34
6. 3-(CH$_2$)—3-C$_5$H$_4$N	3.40
7. 3-O(CH$_2$)$_3$OC$_6$H$_4$—2-NHCOC$_6$H$_3$—2-OCH$_3$—5-SO$_2$F	3.85
8. 3-O(CH$_2$)$_3$OC$_6$H$_4$—3-NO$_2$	3.89
9. 3-O(CH$_2$)$_3$OC$_6$H$_4$—4-NHCOC$_6$H$_4$—3-NO$_2$	4.21
10. 3-O(CH$_2$)$_3$OC$_6$H$_4$—2-NHCONHC$_6$H$_4$—4-SO$_2$F	4.22
11. 3-O(CH$_2$)$_3$OC$_6$H$_4$—4-C$_6$H$_5$	4.35
12. 3-O(CH$_2$)$_2$OC$_6$H$_4$—3-NHCOC$_6$H$_4$—3-SO$_2$F	4.37
13. 3-O(CH$_2$)$_3$OC$_6$H$_4$—3-NHCOC$_6$H$_4$—4-SO$_2$F	5.21

a From Baker and Corey [4].

With physical properties as regression variables, Hansch and Yoshimoto found it necessary to introduce three dummy indicator variables for certain specified ring and bridge features to achieve a comparable level of correlation: $r = 0.935$, $s = 0.258$ [5].

By introducing circuit terms, as described in Chapter 3, corresponding to the features suggested by Hansch and Yoshimoto, we find a two-variable equation that gives the same quality correlation as found by Hansch and Yoshimoto for a four-variable equation:

$$\log 1/C = 0.174 \, {}^1\chi^v + 15.749 \, {}^6\chi^v_{CH} + 1.937$$

$$r = 0.935, \qquad s = 0.257, \qquad N = 108$$

The structure of the general class of benzamidines is given in Table III along with some sample data.

The quality of these results on such a complex system suggests the potential for application of molecule connectivity to drug studies.

C.   Total Molecular Orbital Energy and χ

The total energy of a molecule is a resultant of all attractions and repulsions between nuclei and electrons. Semiempirical, all valence electron molecular orbital calculations generate a total molecular energy presumed to behave in a parallel fashion to the actual total energy. The prediction of conformational preference has been approached using this computed value.

From our work to date on molecular connectivity, we would suspect that for the relatively nonpolar hydrocarbons the total energy may be strongly influenced by the topology or connectivity of the molecules. This has indeed proven to be the case. Using $^1\chi$ and $^3\chi_P$ the equation for alkanes butane through the octanes, where n is the number of carbons, is

$$E_{(CNDO/2)} = 0.0293 \, ^1\chi - 0.0070 \, ^3\chi_P + 8.675n + 1.434$$

$$r = >0.9999, \qquad s = 0.00089, \qquad N = 36$$

The calculated values for heptane isomers are shown in Table IV.

This interesting result strongly implicates the role of connectivity in influencing forces governing the computed energy. It is also a strong indication that these two terms of χ are very closely reflecting salient structural features contributing to the value of this property.

The possibilities opened up by this finding should be explored. The influence of heteroatoms on this result should also be investigated. At the level of hydrocarbons, at any rate, this result represents a close union of molecular connectivity and molecular orbital theory.

TABLE IV

Total CNDO/2 Energy and Connectivity Functions

Molecule	Calc E (a.u.) CNDO/2	Calc E (a.u.) ($^1\chi$, $^3\chi$)
Heptane	62.2502	62.2496
2-Methyl hexane	62.2464	62.2459
3-Methyl hexane	62.2448	62.2447
2,3-Dimethyl pentane	62.2398	62.2388
2,4-Dimethyl pentane	62.2425	62.2430
2,2-Dimethyl pentane	62.2398	62.2405
3,3-Dimethyl pentane	62.2366	62.2361
3-Ethyl pentane	62.2428	62.2440
2,2,3-Trimethyl butane	62.2331	62.2322

D. Resonance Energy and χ

The resonance energy of a molecule is defined as the difference between the experimental heats of hydrogenation of unsaturated molecules and the values predicted for imaginary reference molecules with well-defined, noninteracting double and single bonds. A number of schemes have been developed to compute the probable heats of hydrogenation of these reference molecules [6,7].

The resonance energy is assumed to reflect the degree of stability acquired by a molecule when double bonds interact or when pi electrons become delocalized over a conjugated or aromatic structural system.

In our studies on aromatic and unsaturated molecules, we have developed a scheme to compute χ from both a connectivity standpoint as well as a valence structure. It is evident that the greater the difference between $^1\chi$ and $^1\chi^v$, for a hydrocarbon, the greater is the degree of unsaturation. The interesting possibility arises that this difference between χ values may parallel the computed resonance energy.

A good correlation was discovered between $^1\chi - {}^1\chi^v$ and the resonance energy computed by the Klages scheme for a series of aromatic molecules:

$$RE_{(kcal/mole)} = 52.57({}^1\chi - {}^1\chi^v) - 7.92$$

$$r = 0.9972, \qquad s = 3.72, \qquad N = 14$$

The results are shown in Table V.

This close correlation suggests a physical significance for the χ differential and merits further inquiry.

Nearly comparable results have been obtained with a list of 28 benzenoid hydrocarbons for the SCF resonance energies (eV) calculated by Herndon and Ellzey [8], $r = 0.985$, $s = 0.16$. For 11 of these compounds for which the Klages method gives a value, a correlation of 0.977 was obtained. This work will be extended to heteroatomic molecules.

E. Atomic Charges in Alkanes

A fundamental property of a molecule is the electron density associated with each atom. Calculation of dipole moments and suggestions of reactivity are based on such computed charge distributions. Reliable *ab initio* and all valence approximate SCF methods have made such computed atomic charges generally available.

For alkanes it seems reasonable to suppose that the charge distribution is largely governed by the adjacency relations or branching in the carbon skeleton. We have made a preliminary study of this question.

TABLE V

Resonance Energy and Functions of χ

Hydrocarbon	$^1\chi - {}^1\chi^{v\ a}$	Resonance E	
		(kcal) (Klages)[b]	(kcal) (χ)
Biphenyl	1.562	71.0	74.2
Fluorene	1.504	75.9	71.1
1,3,5-Triphenylbenzene	3.018	148.9	150.7
5,12-Dihydronaphthalene	1.925	95.8	93.3
9,10-Dihydroanthracene	1.531	72.2	75.6
9,10-Diphenylanthracene	3.124	151.6	156.3
9,9'-Bianthryl	3.185	166.6	159.5
Diphenylmethane	1.558	67.0	74.0
Toluene	0.816	35.4	35.0
o-Xylene	0.810	35.1	34.7
m-Xylene	0.800	34.8	34.1
p-Xylene	0.800	34.6	34.1
1,2,3-Trimethylbenzene	0.805	34.1	34.4
1,3,5-Trimethylbenzene	0.782	34.3	33.2

[a] Calculated with ring modification terms: −0.5 per ring for $^1\chi$ and −0.333 per ring for $^1\chi^v$.

[b] From Klages [7].

In this present study we have developed an atomic value of chi, $^1\bar{\chi}_i$, computed as follows:

(1) Each bond term c_{ij} is divided equally between the two vertices i and j.

(2) At each vertex i, the half-bond terms $\frac{1}{2}c_{ij}$ are summed for all the attached edges to give $^1\bar{\chi}_i$.

The equation is

$$^1\bar{\chi}_i = \frac{1}{2} \sum_{j=1}^{\delta_i} c_{ij}$$

For alkanes $^1\bar{\chi}_i$ ranges in value from 0.250 for a terminal methyl attached to a vertex of valence 4, as in 2,2,3-trimethylbutane, to a high of 1.000 for the special case of the central atom in neopentane. Quaternary carbons range from 0.85 to 1.000, tertiary from 0.65 to 0.80, and divalent from 0.50 to 0.60. Terminal methyls range from 0.25 to 0.35 (see Fig. 2 for some examples).

Correlation of the number of valence electrons, calculated by the CNDO method for carbon atoms in alkanes, in standard staggered tetrahedral geometries [9] with $^1\bar{\chi}_i$ yielded the result

$$q = 4.044 - 0.1109\ ^1\bar{\chi}_i, \qquad r = 0.972, \quad s = 0.0054, \quad N = 91$$

The symbol q stands for the number of valence electrons on the carbon atom. All the symmetry-independent atoms were used in the series of molecules butane through isomeric octanes. Only four residuals are greater than twice the standard error. Some of the selected results are presented in Table VI.

The correlation presented here deals with the number of valence electrons on each carbon atom. However, the vertex in the graph may be thought of as consisting of the carbon atom together with the attached hydrogen atoms. Thus, we also correlated the total net charge of the carbon atom plus attached hydrogen atoms. The correlation is the same

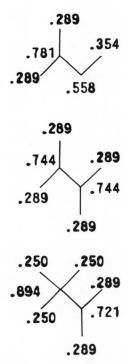

Fig. 2. Atom connectivity values $^1\bar{\chi}$ for three alkane graphs: 2-methylbutane (top), 2,3-dimethylbutane (center), 2,2,3-trimethylbutane (bottom).

TABLE VI

Alkane Carbon Atom Charges and Atomic χ Values

Alkane	$^1\bar{\chi}_i$	q (CNDO/2)	q (Calc)
2-Methylbutane	0.781	3.957	3.958
	0.289	4.012	4.012
	0.289	4.011	4.012
	0.558	3.980	3.982
	0.354	4.013	4.005
2,3-Dimethylbutane	0.744	3.960	3.962
	0.289	4.013	4.012
2,2,3-Trimethylbutane	0.894	3.945	3.945
	0.289	4.012	4.017
	0.721	3.960	3.964
	0.289	4.015	4.012

in quality and the standard error is identical with that for carbon atoms alone.

At this time only one extension of this study has been considered. The atomic values were also computed from $^3\chi_C$. When added as a second variable in a multiple regression, a 10% decrease in the standard error was obtained. The use of other $^m\chi_t$ terms should also be investigated.

The application of this approach to heteroatomic molecules presents some difficulties. The very large local shift in electron density in the vicinity of a nitrogen or oxygen atom, for example, is not explicitly a simple topological phenomenon. It remains to be seen whether some particular use of δ^v values may give adequate description of such electron distributions.

One possible approach to the heteroatomic problem lies in the use of a negative contribution for the edge or subgraph term involving the heteroatom. Such an approach for a carbonyl oxygen, using $\delta^v = 6$, with the present equation leads to a small atomic charge, around -0.08. Thus, an additional variable is required for the simplest possible meaningful application.

VI.   Approaches to the Use of Connectivity in Drug Design

We close our description of molecular connectivity with some general considerations of the application to drug design. In each of the studies discussed in the previous four chapters, we have attempted to show how a calculated correlation may reveal something about the mechanism of action of the drug molecules. We now look at the role that connectivity may play in creating the rationale for new drug design.

A. Interpolation from χ–Activity Equations

The most obvious approach to new compound design, where a quantity of data on a drug molecule series is available, is the use of a calculated structure–activity relationship. Using some of the SAR studies described in the preceding chapters, we can make some predictions about molecules as candidates for synthesis and testing.

1. *Cytochrome P-450 Conversion*

In this study (see Table II, Chapter 7), substituted phenols correlated with activity using $^1\chi^v$ ($r = 0.922$). From the direct relationship calculated, it is predicted that a dibromo derivative should be quite active. For 2,4-dibromophenol, we predict pC $= 2.26$, or about the same activity as 2,4,6-trichlorophenol. Certainly a tribromophenol is predicted to be more active.

2. *Chlorosis in Lemna Minor by Phenols*

In this study (see Table IV, Chapter 9), the Hammett sigma couples with the $^1\chi^v$ term to relate potency to structure ($r = 0.958$). From the equation we predict that extending the alkyl substituent on the ring or introducing a branched alkyl group such as isopropyl would reduce the predicted potency pC.

The equation predicts that a trifluoromethyl group replacing a chloro group would achieve the same predicted potency. In contrast, each bromo replacing a chloro substituent would raise the predicted pC by almost 0.5. Finally, replacement of a methyl group by a nitro group is predicted to enhance the pC by almost 1.0.

3. *General*

This approach constitutes a rationale for substituent replacement that may be chemically feasible and that is predicted to be approximately equipotent with some molecules within the series. Another objective for substituent group modification may be to retain potency while modifying some other biological or physical property.

B. Extrapolations from χ–Activity Equations

An objective of compound design is to maximize the activity of a molecule in a series while retaining desirable physical properties or side effect characteristics. A prediction of a candidate molecule must therefore be an extrapolation from an SAR relationship list while staying below critical values of a chosen physical or side effect property.

1.   *Arenicola Larvae Narcosis*

In Chapter 6, Section IV and Tables III–V, we dissected a loose connectivity–activity relationship of a varied series of molecules into two distinct and well-correlated subclasses. The slope of the subset in Table V is higher than that in Table IV by 30%. The implication is that molecular modification of compounds in Table V would achieve a higher activity than the other list on an atom for atom basis. Thus, if the objective is to minimize the addition of groups that increase molecular weight, the information in the study predicts that a compound in Table V should be the primary candidate for molecular modification.

As an example, from Table IV, the activity of ethyl acetate (M.W. = 88.12) is 0.89, whereas in Table V the activity of toluene (M.W. = 92.15) is 2.29. It would appear, then, that the most fruitful approach to optimum activity is to make synthetic modifications of compounds in the second subclass, Table V.

Alternatively, the intention of the synthetic modification may be to maximize the increase in a physical property that depends closely on χ. If we compare two molecules from each list with comparable χ values, say, ethyl propionate ($^1\chi^v = 2.465$) and toluene ($^1\chi^v = 2.410$), it is apparant that toluene in the Table V list is more active by 50%. Modification of a Table V compound is predicted to lead to a greater rate of activity enhancement starting with approximately the same χ values.

2.   *Tadpole Narcosis*

A similar analysis of a variety of compounds capable of narcotizing tadpoles is revealed in Chapter 6, Table VI. The subclass listed in Table VIII results in a higher slope in the equation relating activity to χ functions. Accordingly it is predicted that molecular modification would be more efficient in increasing activity per atom added, by selecting a molecule from Table VIII.

VII.   Final Summary

In closing we wish to leave the reader with a full measure of encouragement to proceed on his own with investigations studying and using molecular connectivity. After describing the many facets and applications of molecular connectivity developed to this date, we are left with two persistent impressions about it.

The first impression is the extreme simplicity of the computation of the χ functions. It is a process that can be accomplished by a student who has not yet mastered algebra.

The second impression, perhaps more prominent in our minds, is the extensive power that molecular connectivity has in describing characteristics of molecules that are important to many properties. The promise that the approach may shed new light in chemistry is exciting. It is our fond hope that time will reveal even more value for molecular connectivity.

References

1. L. B. Kier, W. J. Murray, M. Randić, and L. H. Hall, *J. Pharm. Sci.* **65** (1976).
2. J. E. Amoore, J. E. Paliere, and E. Wauke, *Nature* **216**, 1084 (1967).
3. R. W. Taft, *in* "Steric Effects in Organic Chemistry" (M. S. Newman, ed.), p. 644. Wiley, New York, 1956.
4. B. R. Baker and M. Cory, *J. Med. Chem.* **14**, 805 (1971).
5. C. Hansch and M. Yoshimoto, *J. Med. Chem.* **17**, 1160 (1974).
6. J. L. Franklin, *Ind. Eng. Chem.* **41**, 1070 (1949).
7. F. Klages, *Chem. Ber.* **82**, 358 (1949).
8. W. C. Herndon and M. L. Ellzey, *J. Am. Chem. Soc.* **96**, 6631 (1974).

APPENDIX A

The purpose of this appendix is to provide tables of $^{m}\chi_{t}$ values for alkanes and $^{m}\chi_{t}^{v}$ values for alkyl substituted benzenes. The reader may examine these to develop a first-hand feeling for the nature of χ terms. Further, a method is presented here for the calculation of $^{1}\chi^{v}$ from the $^{1}\chi$ values for a variety of commonly occurring heteroatoms. Thus, with these tables an investigator may perform studies using $^{1}\chi$ and $^{1}\chi^{v}$ with a minimum of effort.

I. Tables of $^{m}\chi_{t}$ Terms

Table I contains $^{m}\chi_{t}$ terms up to the sixth order ($m = 6$) for graphs corresponding to the alkanes from ethane through all the isomeric nonanes. Path, cluster, and path/cluster terms are included. The alkane symbols have been used by others and are a form of shorthand for the IUPAC name. The rightmost digit in the symbol gives the number of carbon atoms in the longest chain portion of the skeleton. The preceding symbols give the position and the type of group substituted on the carbon chain. For example, the symbol 4 stands for *n*-butane, 2M4 for 2-methylbutane, and 23ME5 for 2-methyl-3-ethylpentane. In Table I the symbol *n* is the number of carbon atoms, that is, the number of vertices in the hydrogen-suppressed graph.

Table II contains $^{m}\chi_{t}^{v}$ terms for 69 alkyl substituted benzenes up to the sixth order for path, cluster, and path/cluster terms. A notation similar to that for alkanes is also used for the benzenes. The symbol 14MM BZ stands for 1,4-dimethylbenzene (*p*-xylene), 1T4 BZ represents *tert*-butylbenzene, 123MMI3 BZ stands for 1,2-dimethyl-3-isopropylbenzene, and 1(1′M)I4 BZ stands for 2,3-dimethyl-1-phenylbutane.

II. Calculation of $^1\chi^v$ Values from $^1\chi$

The first-order terms $^1\chi$ and $^1\chi^v$ are a summation of bond terms c_k. Hence the $^1\chi^v$ term for a heteroatomic molecule with one functional group differs in only one or two c_k values from the $^1\chi$ value for the alkane with the same hydrogen-suppressed graph as the heteroatomic molecule. It is possible to calculate $^1\chi^v$ by a simple manipulation of the $^1\chi$ alkane values.

For example,

$$^1\chi^v \ (n\text{-propanol}) = {}^1\chi \ (n\text{-butane}) - c_{\text{CH}_2-\text{CH}_3} + c_{\text{CH}_2-\text{OH}}$$

or, taking the two c_k values into one term,

$$^1\chi^v = {}^1\chi - S$$

where

$$S = c_{\text{CH}_2-\text{CH}_3} - c_{\text{CH}_2-\text{OH}}$$

To facilitate such calculations, tables of S values have been prepared for a variety of functional groups. Groups affecting one c_k value are given in Table III for —OH, —NH$_2$, —F, —Cl, —Br, —I, and —COOH. Also included in this table is a set of values for N—C bonds in trisubstituted amines under the heading →N. (See Example 7.)

Table IV contains S values for disubstituted oxygen (ethers), nitrogen (disubstituted amines), and sulfur (sulfides). For these cases one must take account of both bonds whose c_k values are changed in converting from $^1\chi$ to $^1\chi^v$. The table entry is the sum for both c_k values. (See Example 4.)

In Table V are the S values for conversion of a benzene to a pyridine derivative. Since two bonds are affected, the connectivity of both atoms adjacent to the nitrogen is taken into account. A similar approach is used to construct Table VI for quinones. In these cases four bonds are involved.

In Tables III–VI, the delta value(s) are given for the atom(s) attached to the heteroatom. In the case of the quinones there are two carbon atoms attached to each of the two oxygen atoms; pairs of delta values are given for Table VI.

III. Sample Calculation of $^1\chi^v$ from Tables I and II Using the S Terms from Tables III–VI

1. Methanol (from ethane, 2)

$$^1\chi^v = {}^1\chi - S = 1.0000 - 0.5528 = 0.4472$$

TABLE I

Connectivity Terms $^m\chi_t$ up to the Sixth Order for Graphs

Alkane symbol	n	$^0\chi$	$^1\chi$	$^2\chi$	$^3\chi_P$	$^4\chi_P$	$^5\chi_P$	$^6\chi_P$
1. 2	2	2.00000	1.00000					
2. 3	3	2.70710	1.41421	0.70710				
3. 4	4	3.41421	1.91421	1.00000	0.50000			
4. 2M3	4	3.57735	1.73205	1.73205	0.00000			
5. 5	5	4.12132	2.41421	1.35355	0.70710	0.35355		
6. 2M4	5	4.28445	2.27005	1.80209	0.81649	0.00000		
7. 22MM3	5	4.50000	2.00000	3.00000	0.00000	0.00000		
8. 6	6	4.82842	2.91421	1.70710	0.95710	0.50000	0.25000	
9. 3M5	6	4.99156	2.80806	1.92166	1.39384	0.28867	0.00000	
10. 2M5	6	4.99156	2.77005	2.18252	0.86602	0.57735	0.00000	
11. 23MM4	6	5.15470	2.64273	2.48803	1.33333	0.00000	0.00000	
12. 22MM4	6	5.20710	2.56066	2.91421	1.06066	0.00000	0.00000	
13. 7	7	5.53553	3.41421	2.06066	1.20710	0.67677	0.35355	0.17677
14. 3E5	7	5.69867	3.34606	2.09077	1.73205	0.86602	0.00000	0.00000
15. 3M6	7	5.69867	3.30806	2.30209	1.47839	0.69692	0.20412	0.00000
16. 2M6	7	5.69867	3.27005	2.53607	1.13502	0.61237	0.40824	0.00000
17. 23MM5	7	5.86180	3.18073	2.62954	1.78202	0.47140	0.00000	0.00000
18. 24MM5	7	5.86180	3.12589	3.02339	0.94280	0.94280	0.00000	0.00000
19. 33MM5	7	5.91421	3.12132	2.87132	1.91421	0.25000	0.00000	0.00000
20. 22MM5	7	5.91421	3.06066	3.31066	1.00000	0.75000	0.00000	0.00000
21. 223MMM4	7	6.07735	2.94337	3.52072	1.73205	0.00000	0.00000	0.00000
22. 8	8	6.24264	3.91421	2.41421	1.45710	0.85355	0.47855	0.25000
23. 3E6	8	6.40577	3.84606	2.47119	1.85162	1.10517	0.40824	0.00000
24. 4M7	8	6.40577	3.80806	2.68252	1.56294	1.12993	0.28867	0.14433
25. 3M7	8	6.40577	3.80806	2.65564	1.74740	0.75671	0.49279	0.14433
26. 2M7	8	6.40577	3.77005	2.88962	1.38502	0.80258	0.43301	0.28867
27. 34MM6	8	6.56891	3.71874	2.77106	2.25930	0.80473	0.16666	0.00000
28. 23ME5	8	6.56891	3.71874	2.82059	1.99156	1.23148	0.00000	0.00000
29. 33ME5	8	6.62132	3.68198	2.87132	2.56066	0.75000	0.00000	0.00000
30. 23MM6	8	6.56891	3.68073	3.00997	1.88208	0.78867	0.33333	0.00000
31. 24MM6	8	6.56891	3.66390	3.14296	1.57069	0.97140	0.33333	0.00000
32. 25MM6	8	6.56891	3.62589	3.36504	1.32136	0.66666	0.66666	0.00000
33. 33MM6	8	6.62132	3.62132	3.26776	1.88388	0.85355	0.17677	0.00000
34. 22MM6	8	6.62132	3.56066	3.66421	1.28033	0.70710	0.53033	0.00000
35. 234MMM5	8	6.73205	3.55341	3.34715	2.10313	0.76980	0.00000	0.00000
36. 233MMM5	8	6.78445	3.50403	3.49683	2.47417	0.40824	0.00000	0.00000
37. 223MMM5	8	6.78445	3.48138	3.67532	2.09077	0.61237	0.00000	0.00000
38. 224MMM5	8	6.78445	3.41650	4.15863	1.02062	1.22474	0.00000	0.00000
39. 2233MMMM4	8	7.00000	3.25000	4.50000	2.25000	0.00000	0.00000	0.00000
40. 9	9	6.94974	4.41421	2.76776	1.70710	1.03033	0.60355	0.33838
41. 4E7	9	7.11288	4.34606	2.85162	1.97119	1.36908	0.69692	0.14433
42. 3E7	9	7.11288	4.34606	2.82475	2.12062	1.18972	0.57735	0.28867
43. 4M8	9	7.11288	4.30806	3.03607	1.83195	1.18972	0.59486	0.20412
44. 3M8	9	7.11288	4.30806	3.00920	1.99740	0.94692	0.53507	0.34846
45. 2M8	9	7.11288	4.27005	3.24318	1.63502	0.97936	0.56751	0.30618
46. 34ME6	9	7.27602	4.25674	2.96210	2.49744	1.42674	0.33333	0.00000
47. 33EE5	9	7.32842	4.24264	2.91421	3.00000	1.50000	0.00000	0.00000
48. 23ME6	9	7.27602	4.21874	3.20102	2.12665	1.37965	0.53745	0.00000
49. 34MM7	9	7.27602	4.21874	3.15149	2.35937	1.14222	0.40236	0.11785
50. 24ME6	9	7.27602	4.20190	3.31207	1.95943	1.22867	0.66666	0.00000
51. 35MM7	9	7.27602	4.20190	3.26254	2.19858	1.02022	0.56903	0.11785
52. 33ME6	9	7.32842	4.18198	3.26776	2.56066	1.20710	0.35355	0.00000
53. 23MM7	9	7.27602	4.18073	3.36352	2.15109	0.85943	0.55767	0.23570
54. 24MM7	9	7.27602	4.16390	3.52339	1.66524	1.41538	0.35355	0.23570
55. 25MM7	9	7.27602	4.16390	3.48461	1.93373	0.82197	0.68688	0.23570

of Alkanes from Ethane through Isomeric Nonanes

Alkane symbol	$^3\chi_C$	$^4\chi_C$	$^5\chi_C$	$^6\chi_C$	$^4\chi_{PC}$	$^5\chi_{PC}$	$^6\chi_{PC}$
1. 2							
2. 3							
3. 4							
4. 2M3	0.57735						
5. 5	0.00000	0.00000					
6. 2M4	0.40824	0.00000	0.00000	0.00000	0.40824		
7. 22MM3	2.00000	0.50000	0.00000	0.00000	0.00000		
8. 6	0.00000	0.00000	0.00000	0.00000	0.00000		
9. 3M5	0.28867	0.00000	0.00000	0.00000	0.57735	0.28867	
10. 2M5	0.40824	0.00000	0.00000	0.00000	0.28867	0.28867	
11. 23MM4	0.66666	0.00000	0.33333	0.00000	1.33333	0.00000	
12. 22MM4	1.56066	0.35355	0.00000	0.00000	1.06066	0.35355	
13. 7	0.00000	0.00000	0.00000	0.00000	0.00000	0.00000	0.00000
14. 3E5	0.20412	0.00000	0.00000	0.00000	0.61237	0.61237	0.20412
15. 3M6	0.28867	0.00000	0.00000	0.00000	0.49279	0.40824	0.20412
16. 2M6	0.40824	0.00000	0.00000	0.00000	0.28867	0.20412	0.20412
17. 23MM5	0.56903	0.00000	0.23570	0.00000	1.27614	0.70710	0.23570
18. 24MM5	0.81649	0.00000	0.00000	0.00000	0.47140	0.94280	0.23570
19. 33MM5	1.20710	0.25000	0.00000	0.00000	1.70710	1.00000	0.25000
20. 22MM5	1.56066	0.35355	0.00000	0.00000	0.75000	1.00000	0.25000
21. 223MMM4	1.65470	0.28867	0.86602	0.28867	2.59807	0.57735	0.00000
22. 8	0.00000	0.00000	0.00000	0.00000	0.00000	0.00000	0.00000
23. 3E6	0.20412	0.00000	0.00000	0.00000	0.55258	0.63713	0.43301
24. 4M7	0.28867	0.00000	0.00000	0.00000	0.40824	0.55258	0.28867
25. 3M7	0.28867	0.00000	0.00000	0.00000	0.49279	0.34846	0.28867
26. 2M7	0.40824	0.00000	0.00000	0.00000	0.28867	0.20412	0.14433
27. 34MM6	0.47140	0.00000	0.16666	0.00000	1.27614	1.13807	0.66666
28. 23ME5	0.50000	0.00000	0.16666	0.00000	1.13807	1.30473	0.66666
29. 33ME5	0.92677	0.17677	0.00000	0.00000	2.03033	1.81066	0.70710
30. 23MM6	0.56903	0.00000	0.23570	0.00000	1.20710	0.66666	0.66666
31. 24MM6	0.69692	0.00000	0.00000	0.00000	0.69104	0.90236	0.66666
32. 25MM6	0.81649	0.00000	0.00000	0.00000	0.57735	0.33333	0.66666
33. 33MM6	1.20710	0.25000	0.00000	0.00000	1.45710	1.38388	0.70710
34. 22MM6	1.56066	0.35355	0.00000	0.00000	0.75000	0.78033	0.70710
35. 234MMM5	0.85911	0.00000	0.38490	0.00000	1.82136	1.53960	0.96225
36. 233MMM5	1.33915	0.20412	0.69692	0.20412	2.93712	1.63299	0.81649
37. 223MMM5	1.57014	0.28867	0.61237	0.20412	2.29489	1.71754	0.81649
38. 224MMM5	1.96890	0.35355	0.00000	0.00000	0.81649	2.04124	1.02062
39. 2233MMMM4	2.50000	0.50000	2.25000	1.50000	4.50000	1.50000	0.00000
40. 9	0.00000	0.00000	0.00000	0.00000	0.00000	0.00000	0.00000
41. 4E7	0.20412	0.00000	0.00000	0.00000	0.49279	0.67941	0.59486
42. 3E7	0.20412	0.00000	0.00000	0.00000	0.55258	0.59486	0.45052
43. 4M8	0.28867	0.00000	0.00000	0.00000	0.40824	0.49279	0.39073
44. 3M8	0.28867	0.00000	0.00000	0.00000	0.49279	0.34846	0.24639
45. 2M8	0.40824	0.00000	0.00000	0.00000	0.28867	0.20412	0.14433
46. 34ME6	0.40236	0.00000	0.11785	0.00000	1.18688	1.52022	1.20710
47. 33EE5	0.70710	0.12500	0.00000	0.00000	2.12132	2.62132	1.45710
48. 23ME6	0.50000	0.00000	0.16666	0.00000	1.08925	1.20710	1.04044
49. 34MM7	0.47140	0.00000	0.16666	0.00000	1.20710	1.13807	0.92258
50. 24ME6	0.61237	0.00000	0.00000	0.00000	0.76180	1.00886	1.04044
51. 35MM7	0.57735	0.00000	0.00000	0.00000	0.91068	0.90236	0.92258
52. 33ME6	0.92677	0.17677	0.00000	0.00000	1.83210	1.98743	1.28033
53. 23MM7	0.56903	0.00000	0.23570	0.00000	1.20710	0.61785	0.63807
54. 24MM7	0.69692	0.00000	0.00000	0.00000	0.60649	1.05767	0.63807
55. 25MM7	0.69692	0.00000	0.00000	0.00000	0.78147	0.48864	0.63807

TABLE I

Alkane symbol	n	$^0\chi$	$^1\chi$	$^2\chi$	$^3\chi_P$	$^4\chi_P$	$^5\chi_P$	$^6\chi_P$
56. 26MM7	9	7.27602	4.12589	3.71859	1.56294	0.93434	0.47140	0.47140
57. 44MM7	9	7.32842	4.12132	3.66421	1.85355	1.47855	0.25000	0.12500
58. 33MM7	9	7.32842	4.12132	3.62132	2.16421	0.83210	0.60355	0.12500
59. 234MEM5	9	7.43915	4.09142	3.56013	2.18401	1.71260	0.00000	0.00000
60. 234MMM6	9	7.43915	4.09142	3.48866	2.59308	1.02885	0.27216	0.00000
61. 233MME5	9	7.49156	4.06469	3.51583	3.00920	1.06649	0.00000	0.00000
62. 22MM7	9	7.32842	4.06066	4.01776	1.53033	0.90533	0.50000	0.37500
63. 334MMM6	9	7.49156	4.04204	3.65143	2.85766	0.90104	0.14433	0.00000
64. 235MMM6	9	7.43915	4.03658	3.85084	1.98126	1.01573	0.54433	0.00000
65. 223MME5	9	7.49156	4.01938	3.87944	2.21034	1.51342	0.00000	0.00000
66. 233MMM6	9	7.49156	4.00403	3.89328	2.45728	0.93301	0.28867	0.00000
67. 223MMM6	9	7.49156	3.98138	4.05574	2.20008	0.86602	0.43301	0.00000
68. 244MMM6	9	7.49156	3.97716	4.11573	1.91794	1.24950	0.28867	0.00000
69. 224MMM6	9	7.49156	3.95450	4.27820	1.65775	1.18972	0.43301	0.00000
70. 225MMM6	9	7.49156	3.91650	4.49318	1.47168	0.72168	0.86602	0.00000
71. 2334MMMM5	9	7.65470	3.88675	4.13076	2.97606	0.66666	0.00000	0.00000
72. 2234MMMM5	9	7.65470	3.85405	4.39781	2.36602	1.00000	0.00000	0.00000
73. 2233MMMM5	9	7.70710	3.81066	4.48743	2.91421	0.53033	0.00000	0.00000
74. 2244MMMM5	9	7.70710	3.70710	5.29809	1.06066	1.59099	0.00000	0.00000

TABLE II

Connectivity Terms $^m\chi_t^v$ up to the Sixth Order

Alkyl benzene symbol	n	$^0\chi^v$	$^1\chi^v$	$^2\chi^v$	$^3\chi_P^v$	$^4\chi_P^v$	$^5\chi_P^v$	$^6\chi_P^v$
1. BZ	6	3.46410	2.00000	1.15470	0.66666	0.38490	0.22222	0.00000
2. 1M BZ	7	4.38675	2.41068	1.65470	0.94045	0.53437	0.30356	0.06415
3. 14MM BZ	8	5.30940	2.82136	2.15470	1.21823	0.63689	0.52578	0.11111
4. 13MM BZ	8	5.30940	2.82136	2.15815	1.17357	0.80701	0.35911	0.15922
5. 12MM BZ	8	5.30940	2.82735	2.08425	1.42556	0.66267	0.37400	0.11111
6. 1E BZ	8	5.09385	2.97134	1.83915	1.25107	0.71371	0.40710	0.12392
7. 135MMM BZ	9	6.23205	3.23205	2.66506	1.36602	1.20235	0.39433	0.26933
8. 124MMM BZ	9	6.23205	3.23803	2.58771	1.66241	0.89134	0.56273	0.18600
9. 123MMM BZ	9	6.23205	3.24401	2.51680	1.87546	0.89779	0.42012	0.19245
10. 113 BZ	9	5.96410	3.35405	2.56538	1.46623	0.83793	0.47882	0.16533
11. 14EM BZ	9	6.01650	3.38202	2.33915	1.52885	0.82378	0.56663	0.28073
12. 13EM BZ	9	6.01650	3.38202	2.34260	1.49073	0.93206	0.55996	0.19690
13. 12EM BZ	9	6.01650	3.38801	2.28003	1.64215	1.00677	0.47060	0.16814
14. 13 BZ	9	5.80096	3.47134	2.23559	1.38149	0.93335	0.53390	0.19714
15. 1235MMMM BZ	10	7.15470	3.65470	3.02371	2.07136	1.25218	0.58034	0.28050
16. 1234MMMM BZ	10	7.15470	3.66068	2.94935	2.32884	1.09150	0.59300	0.24999
17. 1T4 BZ	10	6.88675	3.66068	3.61602	1.63981	0.93815	0.53668	0.19874
18. 14M13 BZ	10	6.88675	3.76474	3.06538	1.74401	0.95135	0.60712	0.39119
19. 12M13 BZ	10	6.88675	3.76474	3.06883	1.70879	1.02923	0.69245	0.22651
20. 12M13 BZ	10	6.88675	3.77072	3.01128	1.81046	1.23457	0.53752	0.20733
21. 135MME BZ	10	6.93915	3.79271	2.84951	1.68972	1.27310	0.69190	0.28976
22. 124MEM BZ	10	6.93915	3.79869	2.78349	1.88553	1.18115	0.75634	0.22319
23. 124MME BZ	10	6.93915	3.79869	2.77216	1.97956	1.02343	0.70479	0.32859
24. 134MME BZ	10	6.93915	3.79869	2.78349	1.87899	1.24249	0.60084	0.34535
25. 123MEM BZ	10	6.93915	3.80467	2.72391	1.99801	1.40665	0.50978	0.24639

—Continued

	Alkane symbol	$^3\chi_C$	$^4\chi_C$	$^5\chi_C$	$^6\chi_C$	$^4\chi_{PC}$	$^5\chi_{PC}$	$^6\chi_{PC}$
56.	26MM7	0.81649	0.00000	0.00000	0.00000	0.57735	0.40824	0.23570
57.	44MM7	1.20710	0.25000	0.00000	0.00000	1.20710	1.81066	0.97855
58.	33MM7	1.20710	0.25000	0.00000	0.00000	1.45710	1.20710	0.97855
59.	234MEM5	0.80274	0.00000	0.27216	0.00000	1.53671	2.32986	1.55327
60.	234MMM6	0.76148	0.00000	0.32853	0.00000	1.84670	1.76563	1.55327
61.	233MME5	1.09126	0.14433	0.55258	0.14433	2.99901	2.77378	1.70728
62.	22MM7	1.56066	0.35355	0.00000	0.00000	0.75000	0.78033	0.55177
63.	334MMM6	1.25460	0.20412	0.49279	0.14433	2.70823	2.43923	1.70728
64.	235MMM6	0.97728	0.00000	0.23570	0.00000	1.41222	1.05219	1.36082
65.	223MME5	1.51036	0.28867	0.43301	0.14433	1.94643	2.64335	1.70728
66.	233MMM6	1.33915	0.20412	0.69692	0.20412	2.71399	1.81294	1.44337
67.	223MMM6	1.57014	0.28867	0.61237	0.20412	2.23510	1.50316	1.44337
68.	244MM6	1.61535	0.25000	0.00000	0.00000	1.55047	2.22119	1.64749
69.	224MMM6	1.84933	0.35355	0.00000	0.00000	1.04538	1.82685	1.64749
70.	225MMM6	1.96890	0.35355	0.00000	0.00000	1.03867	0.82735	1.44337
71.	2334MMMM5	1.48803	0.16666	1.24401	0.33333	3.97606	2.66666	2.16666
72.	2234MMMM5	1.86602	0.28867	0.66666	0.16666	2.69935	2.95534	2.33333
73.	2233MMMM5	2.20710	0.42677	1.81066	1.13388	4.54809	2.97487	1.76776
74.	2244MMMM5	3.12132	0.70710	0.00000	0.00000	1.06066	3.53553	2.65165

for Graphs of Alkyl Substituted Benzenes

	Alkyl benzene symbol	$^3\chi_C^F$	$^4\chi_C^F$	$^5\chi_P^F$	$^6\chi_C^F$	$^4\chi_{PC}^F$	$^5\chi_{PC}^F$	$^6\chi_{PC}^F$	$^6\chi_{CH}^F$
1.	BZ	0.00000	0.00000	0.00000	0.00000	0.00000	0.00000	0.00000	0.03703
2.	1M BZ	0.16666	0.00000	0.00000	0.00000	0.19245	0.16666	0.12830	0.03207
3.	14MM BZ	0.33333	0.00000	0.00000	0.00000	0.38490	0.30356	0.41467	0.02777
4.	13MM BZ	0.33333	0.00000	0.00000	0.00000	0.35911	0.47022	0.36655	0.02777
5.	12MM BZ	0.28867	0.00000	0.08333	0.00000	0.62200	0.45534	0.41467	0.02777
6.	1E BZ	0.11785	0.00000	0.00000	0.00000	0.25393	0.25393	0.20857	0.03207
7.	135MMM BZ	0.50000	0.00000	0.00000	0.00000	0.50000	0.91367	0.68301	0.02405
8.	124MMM BZ	0.45534	0.00000	0.08333	0.00000	0.79040	0.71178	0.83851	0.02405
9.	123MMM BZ	0.41367	0.00000	0.14433	0.00000	0.99401	0.96823	0.90423	0.02405
10.	113 BZ	0.38490	0.00000	0.09622	0.00000	0.63689	0.51089	0.48874	0.03207
11.	14EM BZ	0.28451	0.00000	0.00000	0.00000	0.44638	0.39518	0.43308	0.02777
12.	13EM BZ	0.28451	0.00000	0.00000	0.00000	0.42437	0.50392	0.51691	0.02777
13.	12EM BZ	0.24639	0.00000	0.05892	0.00000	0.59070	0.68622	0.60460	0.02777
14.	13 BZ	0.11785	0.00000	0.00000	0.00000	0.21941	0.29740	0.27028	0.03207
15.	1235MMMM BZ	0.58034	0.00000	0.14433	0.00000	1.13835	1.34668	1.44118	0.02083
16.	1234MMMM BZ	0.53867	0.00000	0.20683	0.00000	1.36901	1.42184	1.57735	0.02083
17.	1T4 BZ	1.33333	0.25000	0.25000	0.08333	1.21225	1.16068	1.05816	0.03207
18.	14M13 BZ	0.55156	0.00000	0.09622	0.00000	0.82934	0.65408	0.66896	0.02777
19.	13M13 BZ	0.55156	0.00000	0.09622	0.00000	0.80900	0.72253	0.91697	0.02777
20.	12M13 BZ	0.51634	0.00000	0.13144	0.00000	0.90722	1.17357	1.10827	0.02777
21.	135MME BZ	0.45118	0.00000	0.00000	0.00000	0.56903	0.89350	0.91103	0.02405
22.	124MEM BZ	0.41306	0.00000	0.05892	0.00000	0.76236	0.89627	1.07177	0.02405
23.	124MME BZ	0.40652	0.00000	0.08333	0.00000	0.85566	0.75282	0.90745	0.02405
24.	134MME BZ	0.41306	0.00000	0.05892	0.00000	0.75909	0.94971	0.94961	0.02405
25.	123MEM BZ	0.37706	0.00000	0.10206	0.00000	0.88235	1.28688	1.23112	0.02405

TABLE II

Alkyl benzene symbol	n	$^0\chi^v$	$^1\chi^v$	$^2\chi^v$	$^3\chi^v_P$	$^4\chi^v_P$	$^5\chi^v_P$	$^6\chi^v_P$
26. 123MME BZ	10	6.93915	3.80467	2.71258	2.09771	1.19487	0.60128	0.22774
27. 114 BZ	10	6.67120	3.82718	3.08356	1.48677	1.08549	0.62175	0.24785
28. 13M3 BZ	10	6.72361	3.88202	2.73905	1.62115	1.15632	0.64838	0.33892
29. 14M3 BZ	10	6.72361	3.88202	2.73559	1.65927	1.04342	0.69878	0.30961
30. 12M3 BZ	10	6.72361	3.88801	2.67648	1.78059	1.15992	0.71392	0.23645
31. 1S4 BZ	10	6.67120	3.89206	2.71997	1.98098	1.01727	0.58236	0.22511
32. 14EE BZ	10	6.72361	3.94268	2.52360	1.83946	1.01066	0.62178	0.38131
33. 13EE BZ	10	6.72361	3.94268	2.52705	1.80788	1.06949	0.70102	0.31087
34. 12EE BZ	10	6.72361	3.94867	2.47581	1.88018	1.24732	0.69220	0.22517
35. 13MT4 BZ	11	7.80940	4.07136	4.11947	1.88409	1.11157	0.79684	0.25172
36. 14MT4 BZ	11	7.80940	4.07136	4.11602	1.91759	1.05356	0.64433	0.47767
37. 124MM13 BZ	11	7.80940	4.18140	3.49839	2.19763	1.12373	0.80801	0.42384
38. 124M13M BZ	11	7.80940	4.18140	3.51474	2.05674	1.38190	0.88390	0.25172
39. 123M13M BZ	11	7.80940	4.18739	3.46018	2.11947	1.73803	0.57190	0.28322
40. 123MM13 BZ	11	7.80940	4.18739	3.44383	2.26853	1.39925	0.72095	0.25544
41. 1T5 BZ	11	7.59385	4.22134	3.60346	2.43914	1.09346	0.62634	0.25051
42. 13M14 BZ	11	7.59385	4.23786	3.58702	1.72643	1.31052	0.71709	0.43260
43. 14M14 BZ	11	7.59385	4.23786	3.58356	1.76455	1.19556	0.78899	0.33824
44. 12M14 BZ	11	7.59385	4.24385	3.52445	1.88941	1.27894	0.87500	0.28377
45. 1(1′M)14 BZ	11	7.54145	4.26474	3.44337	2.32535	1.14150	0.65408	0.26652
46. 124M3M BZ	11	7.64626	4.29869	3.17993	2.02397	1.33893	0.96127	0.36010
47. 134MM3 BZ	11	7.64626	4.29869	3.17993	2.01743	1.39564	0.84914	0.37230
48. 14MS4 BZ	11	7.59385	4.30274	3.21997	2.25875	1.13068	0.71502	0.41477
49. 13MS4 BZ	11	7.59385	4.30274	3.22343	2.22353	1.21234	0.76464	0.34247
50. 12MS4 BZ	11	7.59385	4.30873	3.16588	2.33175	1.35962	0.73618	0.26311
51. 14E13 BZ	11	7.59385	4.32540	3.24983	2.05463	1.13823	0.66860	0.45463
52. 13E13 BZ	11	7.59385	4.32540	3.25328	2.02595	1.17215	0.80134	0.38800
53. 115 BZ	11	7.37831	4.32718	3.41812	1.85318	1.10001	0.79680	0.34892
54. 12E13 BZ	11	7.59385	4.33138	3.20706	2.05800	1.41941	0.83825	0.26436
55. 135MME BZ	11	7.64626	4.35337	3.03396	2.01342	1.35624	0.92968	0.38593
56. 124MEE BZ	11	7.64626	4.35935	2.96794	2.20269	1.32563	0.83862	0.44152
57. 124EEM BZ	11	7.64626	4.35935	2.97927	2.12357	1.42874	0.91946	0.38254
58. 123EEM BZ	11	7.64626	4.36533	2.91969	2.24171	1.60018	0.81595	0.28169
59. 13M4 BZ	11	7.43072	4.38202	3.09260	1.90148	1.24855	0.80696	0.40144
60. 14M4 BZ	11	7.43072	4.38202	3.08915	1.93960	1.13564	0.85409	0.40306
61. 1S5 BZ	11	7.37831	4.39206	3.10040	2.09029	1.38125	0.70917	0.29832
62. 1(1′E)3 BZ	11	7.37831	4.43006	2.92410	2.25657	1.48528	0.68590	0.28489
63. 13E3 BZ	11	7.43072	4.44268	2.92350	1.93831	1.29375	0.79819	0.41061
64. 12E3 BZ	11	7.43072	4.44867	2.87226	2.01862	1.41563	0.86230	0.38187
65. 15 BZ	11	7.21517	4.47134	2.94270	1.91182	1.22380	0.75443	0.39662
66. 131313 BZ	12	8.46410	4.70811	3.97952	2.24401	1.27724	0.88490	0.49485
67. 121313 BZ	12	8.46410	4.71410	3.93831	2.24002	1.56246	1.03437	0.30356
68. 1S6 BZ	12	8.08542	4.89206	3.45395	2.35929	1.45854	0.96654	0.38799
69. 16 BZ	12	7.92228	4.97134	3.29625	2.16182	1.40057	0.89459	0.44274

—Continued

	Alkyl benzene symbol	$^3\chi_C^v$	$^4\chi_C^v$	$^5\chi_C^v$	$^6\chi_C^v$	$^4\chi_{PC}^v$	$^5\chi_{PC}^v$	$^6\chi_{PC}^v$	$^6\chi_{CH}^v$
26.	123MME BZ	0.37139	0.00000	0.12319	0.00000	0.97164	1.10570	1.19084	0.02405
27.	114 BZ	0.52609	0.00000	0.00000	0.00000	0.40824	0.56820	0.52002	0.03207
28.	13M3 BZ	0.28451	0.00000	0.00000	0.00000	0.38985	0.55006	0.54074	0.02777
29.	14M3 BZ	0.28451	0.00000	0.00000	0.00000	0.41186	0.43866	0.49787	0.02777
30.	12M3 BZ	0.24639	0.00000	0.05892	0.00000	0.56081	0.66409	0.75060	0.02777
31.	1S4 BZ	0.30034	0.00000	0.06804	0.00000	0.71520	0.72572	0.67816	0.03207
32.	14EE BZ	0.23570	0.00000	0.00000	0.00000	0.50786	0.48681	0.46801	0.02777
33.	13EE BZ	0.23570	0.00000	0.00000	0.00000	0.48963	0.55191	0.61062	0.02777
34.	12EE BZ	0.20412	0.00000	0.04166	0.00000	0.58416	0.81183	0.85182	0.02777
35.	13MT4 BZ	1.50000	0.25000	0.25000	0.08333	1.38536	1.33552	1.62718	0.02777
36.	14MT4 BZ	1.50000	0.25000	0.25000	0.08333	1.40470	1.30502	1.19590	0.02777
37.	124MM13 BZ	0.67357	0.00000	0.17955	0.00000	1.24029	0.97467	1.25418	0.02405
38.	124M13M BZ	0.68301	0.00000	0.13144	0.00000	1.08034	1.34868	1.68174	0.02405
39.	123M13M BZ	0.64951	0.00000	0.15550	0.00000	1.13835	1.97514	1.96125	0.02405
40.	123MM13 BZ	0.64134	0.00000	0.19716	0.00000	1.29212	1.53641	1.81691	0.02405
41.	1T5 BZ	1.04044	0.17677	0.20118	0.05892	1.72584	1.59426	1.49792	0.03207
42.	13M14 BZ	0.69276	0.00000	0.00000	0.00000	0.57869	0.82204	0.76336	0.02777
43.	14M14 BZ	0.69276	0.00000	0.00000	0.00000	0.60069	0.70945	0.74898	0.02777
44.	12M14 BZ	0.65464	0.00000	0.05892	0.00000	0.75169	0.88790	1.14648	0.02777
45.	1(1'M)14 BZ	0.59622	0.00000	0.22222	0.00000	1.28867	1.07105	1.17490	0.03207
46.	124M3M BZ	0.41306	0.00000	0.05892	0.00000	0.73247	0.87645	1.18497	0.02405
47.	134MM3 BZ	0.41306	0.00000	0.05892	0.00000	0.72920	0.92758	1.10059	0.02405
48.	14MS4 BZ	0.46701	0.00000	0.06804	0.00000	0.90765	0.86891	0.86526	0.02777
49.	13MS4 BZ	0.46701	0.00000	0.06804	0.00000	0.88731	0.94331	1.04630	0.02777
50.	12MS4 BZ	0.43179	0.00000	0.10703	0.00000	0.99585	1.28054	1.40807	0.02777
51.	14E13 BZ	0.50275	0.00000	0.09622	0.00000	0.89082	0.74571	0.71120	0.02777
52.	13E13 BZ	0.50275	0.00000	0.09622	0.00000	0.87426	0.77685	0.95461	0.02777
53.	115 BZ	0.52609	0.00000	0.00000	0.00000	0.50809	0.40652	0.49250	0.03207
54.	12E13 BZ	0.47407	0.00000	0.11735	0.00000	0.91166	1.19891	1.45170	0.02777
55.	135MME BZ	0.40236	0.00000	0.00000	0.00000	0.63807	0.88763	1.08130	0.02405
56.	124MEE BZ	0.36425	0.00000	0.05892	0.00000	0.82762	0.95065	1.09376	0.02405
57.	124EEM BZ	0.37079	0.00000	0.04166	0.00000	0.75582	1.02797	1.24932	0.02405
58.	123EEM BZ	0.33478	0.00000	0.08711	0.00000	0.88309	1.33685	1.53355	0.02405
59.	13M4 BZ	0.28451	0.00000	0.00000	0.00000	0.38985	0.52566	0.57337	0.02777
60.	14M4 BZ	0.28451	0.00000	0.00000	0.00000	0.41186	0.41425	0.052861	0.02777
61.	1S5 BZ	0.30034	0.00000	0.06804	0.00000	0.65541	0.78110	0.81014	0.03207
62.	1(1'E)3 BZ	0.24056	0.00000	0.04811	0.00000	0.70253	0.96333	0.93374	0.03207
63.	13E3 BZ	0.23570	0.00000	0.00000	0.00000	0.45511	0.59806	0.64455	0.02777
64.	12E3 BZ	0.20412	0.00000	0.04166	0.00000	0.55427	0.80721	0.92843	0.02777
65.	15 BZ	0.11785	0.00000	0.00000	0.00000	0.21941	0.27300	0.28376	0.03207
66.	131313 BZ	0.76980	0.00000	0.19245	0.00000	1.25890	1.00460	1.26005	0.02777
67.	121313 BZ	0.74401	0.00000	0.19444	0.00000	1.24401	1.51780	2.16958	0.02777
68.	1S6 BZ	0.30034	0.00000	0.06804	0.00000	0.65541	0.73882	0.84930	0.03207
69.	16 BZ	0.11785	0.00000	0.00000	0.00000	0.21941	0.27300	0.28276	0.03207

TABLE III

Value Subtracted from ¹χ to Give ¹χᵖ for Monofunctionals: S Terms

δ	—OH	—NH₂	→N	—F	—Cl	—Br	—I	—COOH
1	0.5528	0.4224	0.1301	1.2236	−0.2039	−0.9842	−2.4300	0.8043
2	0.3909	0.2989	0.0920	0.8652	−0.1442	−0.6959	−1.7183	0.7817
3	0.3192	0.2440	0.0751	0.7065	−0.1177	−0.5682	−1.4030	0.7716
4	0.2764	0.2113	0.0651	0.6118	−0.1019	−0.4921	−1.2150	0.7657

TABLE IV

Value Subtracted from ¹χ to Give ¹χᵖ for Disubstituted Nitrogen, Oxygen, and Sulfur: S Terms

	Disubstituted											
	nitrogen				oxygen				sulfur			
δ	1	2	3	4	1	2	3	4	1	2	3	4
1	0.5198	0.4437	0.4099	0.3898	0.5977	0.5102	0.4714	0.4483	−0.9780	−0.8347	−0.7713	−0.7335
2		0.3675	0.3338	0.3137		0.4227	0.3839	0.3608		−0.6915	−0.6281	−0.5903
3			0.3000	0.2800			0.3451	0.2571			−0.5146	−0.5268
4				0.2599				0.2989				−0.4890

TABLE V

Values Subtracted from $^1\chi$ *to Give* $^1\chi^v$ *for Pyridines: S Terms*

δ	3	4
3	0.1503	0.1402
4		0.1302

TABLE VI

Values Subtracted from $^1\chi$ *to Give* $^1\chi^v$ *for Quinones: S Terms*

δ	3,3	3,4	4,4
3,3	0.3905	0.3774	0.3644
3,4		0.3644	0.3513
4,4			0.3382

TABLE VII

Q Values for Conversion of $^1\chi^v$ *to* $^1\chi$ *for Alkyl Substituted Benzenes*

Substitution pattern	Q
—	1.0000
1	0.9058
12	0.8225
13, 14	0.8116
123	0.7392
124	0.7283
135	0.7174

TABLE VIII

R Values for Conversion of $^1\chi^v$ to $^1\chi$ for Exo Bonds for Alkyl Substituted Benzenes

	1	2	3	4
R	0.0774	0.0545	0.0447	0.0387

2. Propanol (from butane, 4)

$$^1\chi^v = {}^1\chi - S = 1.9142 - 0.3909 = 1.5233$$

3. 2,3-Dimethyl-2-butanol (from 223MMM5)

$$^1\chi^v = {}^1\chi - S = 3.4814 - 0.2764 = 3.2050$$

4. Isopropylethyl ether (from 2M5)

$$^1\chi^v = {}^1\chi - S = 2.7701 - 0.3839 = 2.3862$$

5. 3-Fluoropentane (from 3M5)

$$^1\chi^v = {}^1\chi - S = 2.8081 - 0.7065 = 2.1016$$

6. 1,1,1-Trifluoroethane (from 22MM3)

$$^1\chi^v = {}^1\chi - S = 2.0000 - 3(0.6118) = 0.1646$$

7. Isopropyldimethylamine (from 23MM4). This calculation may be performed by considering each of the three bonds to nitrogen separately, using the S values in Table III listed under $\to N(5)$ as follows:

$$^1\chi^v = {}^1\chi - S_a - S_b - S_c$$
$$= 2.6427 - 0.0751 - 0.1301 - 0.1301 = 2.3074$$

8. Pyridine (from benzene)

$$^1\chi^v = {}^1\chi - S = 2.0000 - 0.1503 = 1.8497$$

9. 1,3-Dimethyl pyridine (from 14MM BZ)

$$^1\chi^v = 2.8214 - 0.1402 = 2.6812$$

10. Bromobenzene (from 1M BZ)

$$^1\chi^v = 2.0000 - (-0.4921) = 2.4921$$

11. *m*-Chlorophenol (from 13MM BZ)

$$^1\chi^v = {}^1\chi - S_{OH} - S_{Cl} = 2.0000 - 0.2764 - (-0.1019) = 1.8255$$

12. 1-Phenyl-3-aminopropanol (from 1(1'M)I4 BZ)

$$^1\chi^v = 4.2647 - S_{OH} - S_{NH_2} = 4.2647 - 0.3192 - 0.2989 = 3.6466$$

IV. Calculation of $^1\chi$ from $^1\chi^v$ for Substituted Benzenes and Related Compounds

Table II can be used to calculate $^1\chi$ values from $^1\chi^v$ values. In the computation of $^1\chi^v$ the δ^v values of 3 or 4 are used for the vertices in the benzene ring; 4 is used where there is a side chain attached. To compute $^1\chi$ from $^1\chi^v$ then, a correction term Q must be developed for each pattern of substitution. These Q values have been given in Table VII for the substitution patterns that occur in the compounds listed in Table II.

In addition, the c_k term for each edge directly attached to the ring must be changed. In the $^1\chi^v$ term the possibilities for delta values in these bonds are 4–1, 4–2, 4–3, 4–4, whereas in $^1\chi$ they are 3–1, 3–2, 3–3, and 3–4. These correction terms R are tabulated in Table VIII.

V. Sample Calculations of $^1\chi$ from $^1\chi^v$

1. Benzene (from BZ)

$$^1\chi = 2.0000 + Q(1) = 2.0000 + 1.0000 = 3.0000$$

2. Bromobenzene, chlorobenzene, phenol, aniline (from 1M BZ)

$$\begin{aligned} ^1\chi &= 2.4107 + Q(1) + R(1) \\ &= 2.4107 + 0.9058 + 0.0774 = 3.394 \end{aligned}$$

3. *m*-Chlorophenol, *m*-xylene, *m*-aminotoluene

$$\begin{aligned} ^1\chi &= 2.8214 + Q(13) + 2R(1) \\ &= 2.8214 + 0.8116 + 2(0.0774) = 3.7878 \end{aligned}$$

4. *p*-Methoxy-*m*-ethylphenol (from 124 EEM BZ)

$$\begin{aligned} ^1\chi &= 4.3594 + Q(124) + 2R(2) + R(1) \\ &= 4.3594 + 0.7283 + 2(0.0545) + 0.0774 = 5.2741 \end{aligned}$$

5. 1,3-Dimethylpyridine (14MM BZ)

$$\begin{aligned} ^1\chi &= 2.8214 + Q(14) + 2R(1) \\ &= 2.8214 + 0.8116 + 2(0.0774) = 3.7878 \end{aligned}$$

APPENDIX B

Table I contains the correlation coefficient r between each of the $^m\chi_t$ terms for the set of graphs of alkanes given in Appendix A, Table I. A large value of r indicates the two terms are not completely independent. A negative value indicates that as one increases in value, the other term decreases. It should be noted that the highest correlation is between n and $^0\chi$, as expected from the definition of $^0\chi$ given in Chapter 3.

TABLE I

Correlation Between the $^m\chi_t$ Terms for Alkanes[a]

	n	$^0\chi$	$^1\chi$	$^2\chi$	$^3\chi_P$	$^4\chi_P$	$^5\chi_P$	$^6\chi_P$	$^3\chi_C$	$^4\chi_C$	$^5\chi_C$	$^6\chi_C$	$^4\chi_{PC}$	$^5\chi_{PC}$	$^6\chi_{PC}$
n	1.000														
$^0\chi$	0.991	1.000													
$^1\chi$	0.978	0.941	1.000												
$^2\chi$	0.800	0.864	0.659	1.000											
$^3\chi_P$	0.800	0.814	0.779	0.569	1.000										
$^4\chi_P$	0.775	0.738	0.811	0.480	0.516	1.000									
$^5\chi_P$	0.502	0.422	0.587	0.197	0.143	0.382	1.000								
$^6\chi_P$	0.317	0.234	0.409	0.043	0.021	0.224	0.558	1.000							
$^3\chi_C$	0.309	0.422	0.111	0.799	0.200	0.041	−0.224	−0.286	1.000						
$^4\chi_C$	0.164	0.261	−0.007	0.630	0.065	−0.042	−0.249	−0.251	0.931	1.000					
$^5\chi_C$	0.185	0.269	0.063	0.418	0.455	−0.204	−0.315	−0.238	0.487	0.381	1.000				
$^6\chi_C$	0.114	0.183	0.063	0.350	0.307	−0.239	−0.242	−0.155	0.468	0.427	0.924	1.000			
$^4\chi_{PC}$	0.415	0.510	0.279	0.602	0.726	0.031	−0.315	−0.334	0.612	0.478	0.861	0.719	1.000		
$^5\chi_{PC}$	0.563	0.644	0.442	0.700	0.660	0.520	−0.279	−0.335	0.652	0.537	0.451	0.329	0.744	1.000	
$^6\chi_{PC}$	0.625	0.689	0.522	0.711	0.616	0.588	0.014	−0.273	0.570	0.424	0.299	0.137	0.593	0.903	1.000

[a] The entries in the table are the correlation coefficient r between each pair of $^m\chi_t$ terms for the values recorded for the alkane graphs in Table I, Appendix A.

INDEX